AF335548

Distinguished Dissertations

Springer

London
Berlin
Heidelberg
New York
Barcelona
Hong Kong
Milan
Paris
Singapore
Tokyo

Russ Bubley

Randomized Algorithms: Approximation, Generation, and Counting

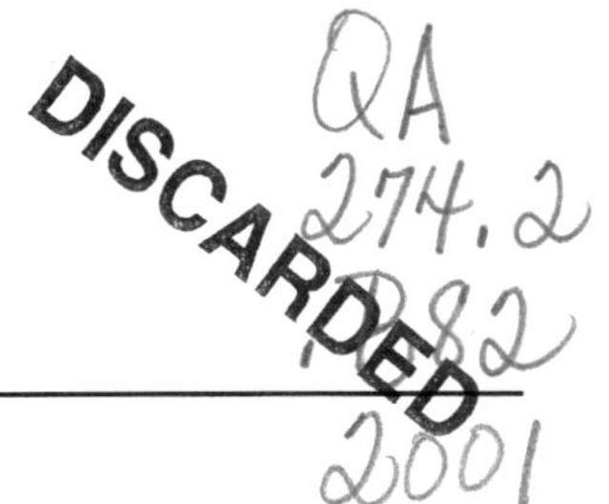

Springer

Russ Bubley, MA, PhD
University of Leeds, Leeds, UK

Series Editor

Professor C.J. van Rijsbergen
Department of Computing Science, University of Glasgow, G12 8RZ, UK

ISBN 1-85233-325-1 Springer-Verlag London Berlin Heidelberg

British Library Cataloguing in Publication Data
Bubley, Russ
    Randomized algorithms : approximation, generation and
    counting. - (Distinguished dissertations)
    1.Probabilities 2.Stochastic approximation 3.Algorithms
    4.Computer science - Mathematics
    I.Title
    519.2
    ISBN 1852333251

Library of Congress Cataloging-in-Publication Data
Bubley, Russ, 1974-
    Randomized algorithms : approximation, generation, and counting / Russ Bubley.
        p. cm. -- (Distinguished dissertaions, ISSN 1439-9768)
    Includes bibliographical references.
    ISBN 1-85533-325-1 (alk. paper)
    1. Markov processes. 2. Monte Carlo method. 3. Computational complexity. I. Title.
    II. Distinguished dissertaions (Springer-Verlag)
    QA274.2 .B82 2000
    519.2'33--dc21   00-034426

Typesetting: Camera-ready by author
Printed and bound by the Athenæum Press Ltd., Gateshead, Tyne & Wear
34/3830-543210 Printed on acid-free paper SPIN 10767866

Dedication

For those closest to me.

Preface

What This Book Is

In this book we concern ourselves primarily with two problems of fine pedigree: *counting* and *generation*. Given a finite set of combinatorial objects, we wish to discover how many objects there are in it, and to pick one uniformly at random. These two operations are of fundamental importance to discrete mathematics and probability.

We also consider the generalization of these problems to the setting where each object has a positive *weight* associated with it. The "counting" problem is then the summation of these weights over all the objects, and the generation (or *sampling*) problem is to sample an object at random with probability proportional to the weights. Indeed in Section 4.3 we consider a further generalization to the case where the set of objects is no longer finite, and is instead some subset of $\mathbb{R}^n$, and the counting problem becomes the integration of the weights, and the generation problem is to sample an element with density proportional to the weights.

The fundamental problem with naïve solutions to these problems is one of time: simple algorithms can take far too long — far longer than we might be prepared to wait. It is normal in computational complexity theory to regard a problem as *tractable* if we know of an algorithm that takes time that is bounded above by some polynomial of the size of the problem description; naïve solutions may easily grow exponentially instead.

Unfortunately, there are many counting problems that are not known to be tractable; indeed, for many counting problems, there is complexity theoretic evidence to suggest strongly that they are, in fact, intractable.

We need therefore to relax our demands in these cases, and ask for only an approximate answer. We must be satisfied to have a counting solution that is an approximation to the true answer. Also, we may be willing to accept the generation of an object from some distribution that is only approximately proportional to the weights — especially if with high probability we would be unable to tell the difference between the two distributions with a tractable algorithm.

We will thus primarily be interested in developing such approximation algorithms for cases in which exact counting is intractable in some appropriate sense — although we may still be interested in other cases, where approximation algorithms merely provide a faster solution than the slower (but tractable) exact solution. Consequently, we prove many results throughout this book to show that certain counting problems are intractable, before illustrating approximation algorithms for them.

Up to this point, we have not described any connexion between these two apparently disparate problems, counting and generation. The reason for this is simple: the most powerful known technique for approximate counting is the Markov chain Monte Carlo method — at its heart it is driven by a requirement for generation — and approximate sampling is as good for this task as exact generation. Conversely, approximate counting can be used as a powerful technique to approximately sample.

How then do we achieve at least approximate generation? There are currently only two known techniques: direct sampling and Markov chain simulation, the latter being far more powerful. Markov chain simulation consists of picking one object to start with, and then trading this object in for a new (random) one — typically one that is in some sense very similar to our original object; we then repeat this "trading in" step many times. If we have taken enough of these "trading in" steps, and have chosen the probability distribution for each such step in the correct manner, then the final object will have been generated approximately from our desired distribution.

It transpires that the main difficulty in the analysis is typically determining just how many steps of the Markov chain simulation constitute "enough". We focus in this book on a classical technique known as *coupling*, which was introduced by Doeblin in the 1930s. This particular technique had until recently been side-lined in the arena of approximate generation, as it was believed to be too difficult to apply. Jerrum [66] wrote

> [The] "coupling" argument ... is very successful in the context of highly symmetric Markov chains ..., but seems very difficult or even impossible to apply to the kind of Markov chains which arise in the analysis of Monte Carlo algorithms.

It transpires that analysis by coupling can in fact be achieved, and we give several non-trivial applications of a "bare-hands" approach to coupling here. We go on to develop a weakening of the coupling method that makes application substantially easier; this we name *path coupling*. We illustrate the utility of this new method by considering a wide variety of applications.

What This Book Is Not

In this book we *do not* consider approximation algorithms for optimization problems; this is a very active area of research and is outside the scope of this work. The interested reader should consult a recent review, such as [9, 11].

Organization of This Book

In Chapter 1, we introduce the mathematical setting within which the rest of this book is set. We look at computational complexity theory, probability — with particular attention paid to Markov chains — and review some definitions from graph theory. We also introduce some basic results which we will use later on in the book.

In Chapter 2, we consider methods for sampling and approximately sampling from a (possibly weighted) set of combinatorial objects. We review numerous analytical techniques for analyzing the Markov chain method, and introduce a new technique into this domain: *Dobrushin's uniqueness criterion*.

In Chapter 3, we consider methods for approximate counting, and focus on utilizing the Markov chain method for sampling as an aid to this end.

In Chapter 4, we consider three new direct applications of the coupling method to analyze Markov chain approximate samplers, and show how two of these sampling algorithms may be used to construct approximate counting algorithms.

In the Intermezzo, we introduce a new technique for applying the coupling method. This we name *path coupling*. We go on to prove several results that make subsequent applications of the path coupling method much simpler.

In Chapter 5, we go on to use the path coupling technique for a wide variety of problems; we prove both results on new Markov chains, and improve upon previous results for Markov chains introduced elsewhere.

In Chapter 6, we consider several directions in which the work of this book could be extended, and offer possible ways in which these extensions might be achieved.

In Appendix A, we give an application of Dobrushin's uniqueness criterion to a Markov chain problem from theoretical computer science and statistical physics.

In Appendix B, we review the current state of the knowledge of a variety of exact and approximate counting problems related to SAT, the progenitor of modern complexity theory problems.

In Appendix C, we prove the equivalence of a natural algorithmic metric, *transposition distance*, to a classical metric, *Spearman's footrule*.

Acknowledgements

Some portions of the work presented in this book have been published elsewhere.

Much of Section 4.3 has appeared in *Random Structures and Algorithms* (volume 12, pages 213–235, 1998), under the title "An Elementary Analysis of a Procedure for Sampling Points in a Convex Body" (co-authored with Martin Dyer and Mark Jerrum). Copyright © 1997 John Wiley & Sons, Inc. An earlier version of this paper appeared as a technical report from the Department of Computer Science, University of Edinburgh (Report ECS-LFCS-96-343, April 1996), under the title "A New Approach to Polynomial-Time Generation of Random Points in Convex Bodies".

Much of Section 4.2 will appear in the *SIAM Journal on Computing* under the title "Graph Orientations with No Sink and an Approximation Algorithm for a Hard Case of #SAT" (co-authored with Martin Dyer). This paper appeared earlier with a similar title in the *Proceedings of the Eighth Annual ACM–SIAM Symposium on Discrete Algorithms* (New Orleans, Louisiana, 5–7 January 1997, pages 248–257). Copyright © 1997 SIAM.

Portions of the Intermezzo and Chapter 5 have appeared in the *Proceedings of the 38th Annual Symposium on Foundations of Computer Science* (Miami Beach, Florida, 20–22 October 1997, pages 223–231), under the title "Path Coupling: A Technique for Proving Rapid Mixing in Markov Chains" (co-authored with Martin Dyer). Copyright © 1997 IEEE. An earlier version of this paper, which also included much of Section 2.3.2, appeared as a technical report from the School of Computer Studies, University of Leeds (Report 97.04, January 1997), under the title "Path Coupling, Dobrushin Uniqueness, and Approximate Counting".

Portions of Section 5.7 appeared in a special issue of *Discrete Mathematics* (volume 201, pages 81–88, 1999) under the title "Faster Random Generation of Linear Extensions" (co-authored with Martin Dyer). Copyright © 1999 Elsevier Science B.V. This paper appeared earlier in the *Proceedings of the Ninth Annual ACM–SIAM Symposium on Discrete Algorithms* (San Francisco, California, 25–27 January 1998, pages 350–354). Copyright © 1998 SIAM. An earlier version of this paper appeared as a technical report from the School of Computer Studies, University of Leeds (Report 97.41, August 1997).

An expanded version of Section 5.10 appeared in the *SIAM Journal on Computing* (volume 29, pages 387–400, 1999) under the title "On Approximately Counting Colourings of Small Degree Graphs" (co-authored with Martin Dyer, Catherine Green-hill, and Mark Jerrum). This paper appeared earlier in the *Proceedings of the Ninth Annual ACM–SIAM Symposium on Discrete Algorithms* (San Francisco, California, 25–27 January 1998, pages 355–363), under the title "Beating the 2Δ Bound for Approximately Counting Colourings: A Computer-Assisted Proof of Rapid Mixing" (co-authored with Martin Dyer and Catherine Greenhill). Copyright © 1998 SIAM.

This research was funded by an EPSRC grant and partially by the ESPRIT project, RAND2.

Contents

List of Tables

List of Figures

Chapter 1

Mathematical Background

In this chapter we review the mathematical setting within which the research in this book is set. We also introduce much of the common notation that is used elsewhere in this book, and review some simple results that we will use in subsequent chapters.

1.1 Computational Complexity

1.1.1 Introduction

The basic elements of computational complexity theory are now well-known, so in this section we shall restrict ourselves to giving a brief review of some standard notation and common complexity classes. We shall then go on to describe some less well-known complexity classes that are relevant to this book, and introduce some tools for dealing with these. We will not descend to the level of formalism that requires consideration of Turing machine (or equivalent) models, as this would obscure the ideas being presented. For such a presentation, the reader is referred to Garey and Johnson [51] for deterministic complexity classes, and Jerrum, Valiant, and Vazirani [71] or Sinclair [111] for randomized complexity classes. Other readable presentations of this material may be found in Welsh [126] and Motwani and Raghavan [96].

1.1.2 Notation for Asymptotics: O, Ω, and Θ

The notion of the worst-case complexity of an algorithm, and the asymptotic "big-oh" notation for it are probably the most well-known elements of computational complexity. We next present the most often used (and abused) notation.

Following Motwani and Raghavan [96] and, in spirit, Knuth [81], let $f(n), g(n) : \mathbb{N} \to \mathbb{R}$. We write:

1. $f(n) = O(g(n))$ if there exist $c \in (0, \infty)$, and $N \in \mathbb{N}$, such that for all $n > N$, we have $f(n) \leq cg(n)$.

2. $f(n) = \Omega(g(n))$ if there exist $c \in (0, \infty)$, and $N \in \mathbb{N}$, such that for all $n > N$, we have $f(n) \geq cg(n)$.

3. $f(n) = \Theta(g(n))$ if $f(n) = O(g(n))$ and $f(n) = \Omega(g(n))$.

4. $f(n) = o(g(n))$ if $\lim_{n \to \infty} f(n)/g(n) = 0$.

We also use the notation $\tilde{O}(n)$, $\tilde{\Omega}(n)$ and $\tilde{\Theta}(n)$ to mean respectively $O(n)$, $\Omega(n)$, and $\Theta(n)$, where factors that are poly-logarithmic in n have been hidden: i.e. where factors that are $o(\log^c n)$, for some constant c, are hidden.

When we refer to an algorithm *running in time* $O(g(n))$ or *being* $O(g(n))$ ($\Omega(g(n))$ or $\Theta(g(n))$ respectively), we mean that if we define $f(n)$ to be the maximum time taken by the algorithm for any input with parameter n, then $f(n) = O(g(n))$ ($f(n) = \Omega(g(n))$ or $f(n) = \Theta(g(n))$ respectively). In particular this *does not* mean that if an algorithm is $\Omega(g(n))$ the algorithm will take at least this long for all inputs with parameter n.

When we say that a *randomized* algorithm runs in time $O(g(n))$ (or $\Theta(g(n))$), we mean that the *expected* time is $O(g(n))$ or $\Theta(g(n))$ respectively.

1.1.3 Complexity Classes

In order to define the structure of problems better, much effort has been turned to classifying computational problems according to how hard they are to solve. This has given rise to an astonishing proliferation of *complexity classes*. We introduce here the complexity classes that are of relevance to this book; more extensive classifications and more formal definitions may be found in standard works on structural complexity theory, e.g. Papadimitriou [97].

The introduction to complexity classes given here avoids the traditional formalism of Turing machines and their equivalents. Such a treatment based on Turing machines may be found in Garey and Johnson [51], and Welsh [126].

We first consider decision problems: i.e. questions with a yes/no answer, e.g. "Does this integer have a non-trivial factor?".

For our purposes, it is simplest to regard a computational problem as a function that maps from problem instances to solutions. Such a function is said to be *polynomial-time computable* if there is an algorithm for computing it that runs in time $O(n^c)$ for some constant, c, where n is the *natural size* of the problem instance.

The concept of *natural size* is discussed in Garey and Johnson [51]. In this book however, we typically make the parameterization of natural size explicit, for clarity.

The class P is the class of decision problems for which there exists a polynomial-time computable function that maps (correctly) from problem instances to yes/no answers.

The class NP is slightly more complicated. Once again it is a class of decision problems. This time, however, we take as input both the problem instance and some evidence that the answer is yes. (This evidence is usually referred to as a *witness*, or a *certificate*.) The evidence must have natural size that is polynomial in the size of the problem instance. Once again, we must have a polynomial-time computable function, this time from the instance/evidence combination to yes/no. This function must be such that if an instance is a 'no' instance, the function always maps to no, whatever the evidence, and if the function is a 'yes' instance, it must map to a yes for at least one piece of evidence.

Note that for an instance/evidence combination that results in a yes under this function, we may interpret the evidence as a short proof that the instance is a 'yes' instance, with the polynomial-time computable function as a proof verifier.

Note that the definition of NP is asymmetric in yes and no, whereas that of P was symmetric. The class of problems that lie in the complement of NP is called co-NP.

The class of decision problems, RP, is somewhat simpler to define. A problem is in RP if there is a polynomial-time algorithm for it that computes a random variable, such that a 'yes' instance is mapped to yes with probability at least $1/2$, a 'no' instance is mapped to no with probability 1.

By analogy with co-NP, co-RP is defined as the complement of RP. Define ZPP $=$ RP $\cap$ co-RP.

A decision problem is in the class BPP if there is a polynomial-time algorithm for it that computes a random variable, such that a 'yes' instance is mapped to yes with probability at least $3/4$, a 'no' instance is mapped to no with probability at least $3/4$.

Remark. The use of $1/2$ in the definition of RP, and of $1/4$ and $3/4$ in the definition of BPP is purely by convention. The $1/2$ could be replaced by an inverse polynomial of the input size, and the $3/4$ and $1/4$ by $1/2$ plus or minus an inverse polynomial of the input size, all without materially affecting the definitions.   ○

Finally, we shall define a class of problems that is not a class of decision problems, but is instead a class of *counting problems*. Recall the definition of NP, where for a 'yes' instance, there must be some piece of evidence which would map (with the problem instance) to yes, under the polynomial-time function. The problem of establishing how many such pieces of evidence there are for a problem instance is the definitive problem of the class #P. Note that this depends not only on the decision problem but also on the type of witness. For example, for the NP problem "Does this number have a non-trivial factor?", a valid witness might be a non-trivial factor, however an equally valid witness might be either the factor itself, or the factor ± 1. Clearly, both of these types of witness would lead to polynomial-time proof verifiers, and the decision problem remains the same, but the #P counting problem is different.[1]

1.1.4   Oracles, Reductions, and Hardness

The concept of an oracle is a very useful one when considering complexity classes. For our purposes, we may consider an oracle for a problem to be a black-box device that solves an instance of the problem in time $O(1)$. Similarly, an oracle for a class of problems is simply a black-box device that can solve any instance of any problem in the class in time $O(1)$.

---

[1] Indeed for this problem there is no obvious polynomial-time function for transforming between the solutions to these two counting problem variants.

Define a Cook–Levin reduction as follows. A problem A is said to be Cook–Levin reducible to a problem B, if problem A may be solved in *deterministic* polynomial time, given access to *one* call to an oracle for B.

Similarly, a Turing (or Karp) reduction may be defined by relaxing the definition of a Cook–Levin reduction to allow polynomially many calls to the oracle.

Remark. Cook–Levin reductions are sometimes referred to simply as Cook reductions, or polynomial reductions. They were introduced along with the concept of NP-complete problems independently by Cook [30] and Levin [84]. $\bigcirc$

A problem is said to be *hard* for a class, if any problem in the class is Turing reducible to it. Hence we obtain the classification of problems as NP-hard or #P-hard.

In general, a problem is said to be *complete* for a class if it is both hard for the class and contained within the class, e.g. #P-complete. For historical reasons, NP-complete has a subtly different definition. For a problem to be NP-complete, it must be in NP, and also, every problem in NP must be Cook–Levin reducible to it.

A problem is said to be *easy* for a class if it is Turing reducible to a complete problem for the class.

1.1.5 More Complexity Classes

It is easy to construct further complexity classes, for recall the definition of NP: a decision problem is in NP if there is a polynomial-time function from instance plus evidence to yes/no. We may construct a hierarchy of classes now that we have the definition of an oracle. Define the class Σ_0^p to be P, and define the class Σ_{k+1}^p to be the set of decision problems for which there is a polynomial-time computable function from instance plus evidence to yes/no, where the function has access to an oracle for Σ_k^p. Thus Σ_1^p is NP.

Together, these classes form the polynomial hierarchy, i.e.

$$\mathrm{PH} = \bigcup_{k=0}^{\infty} \Sigma_k^p,$$

and a problem that is easy for Σ_k^p is said to be at the kth level of the polynomial hierarchy.

Toda [119] has produced a very strong result relating PH and #P. Specifically, he has shown that problems that are #P-complete are Σ_k^p-hard, for all k.

1.1.6 Conclusion

We have introduced all of the complexity classes that we shall deal with in this book, without giving any intuition as to how these classify problems in terms of computational ease of solubility.

Since there are no known sub-exponential time algorithms for NP-complete problems, we shall say that all problems that are BPP-easy are *easy*, or *tractable*[2] problems.

---

[2]Note that this definition of *tractable* is more general than is sometimes used, as it allows randomization.

We shall make the assumption that no NP-complete problems are tractable, and may refer to problems that are NP-hard as being intractable. We shall also assume, without making further explicit reference, that $\text{RP} \subsetneq \text{NP}$.

1.1.7 Some Common Problems in Complexity Theory

In this section, we list a few common problems that are normally referred to in what to the uninitiated may appear to be an appallingly obscure manner. In order to alleviate this burden, we provide a list here of some of the problems to which we refer in later chapters. In order to avoid ambiguities of the type discussed in Section 1.1.3, we present these in the form of counting problems: the decision problem variants are simply the questions "Does the counting problem have a result of at least one?".

- Graph k-Colouring:

    Given a graph on vertex set V, how many mappings are there from V to a set of colours of size k, such that no two adjacent vertices are mapped to the same colour?

    The decision problem is NP-complete, see Garey, Johnson, and Stockmeyer [52]. That the counting problem is #P-complete follows from work of Jaeger, Vertigan, and Welsh [65].

- Permanent/Perfect Matchings:

    Given an $n \times n$ $(0, 1)$-matrix, A, compute the permanent of A:

    $$\text{Per}(A) = \sum_{\sigma} \prod_{i=1}^{n} A_{i,\sigma(i)},$$

    where the sum is over all permutations of $\{1, 2, \ldots, n\}$.

    If the matrix A is regarded as the adjacency matrix of a bipartite graph, then the problem may be restated as "How many perfect matchings (see Section 1.4) does the graph have?".

    The decision problem may be computed in polynomial-time, but the counting problem is #P-complete [124, 13].

- SAT

    Given a Boolean expression in conjunctive normal form, how many mappings are there from the set of variables in the expression to $\{\text{True}, \text{False}\}$, such that the entire expression is logically equal to True?

    The decision problem is NP-complete, see Cook [30]. The counting problem is #P-complete, a result of Valiant [125].

    A variant of this problem is DNF, in which the Boolean expression is given in disjunctive normal form. The decision problem is then trivially polynomial-time, but the counting problem remains #P-complete.

1.2 Probability

As one of our primary motivations in this book is to sample approximately from some probability distribution, we must be able to formalize what we mean by "approximately" in this context. We will say that one distribution approximates another with respect to some norm.

For our purposes, we will typically use the *total variation distance norm* [58] (or simply *variation distance*), and write this as $d_{TV}(\cdot, \cdot)$. We use this norm as it dovetails naturally with our primary tool for analyzing Markov chains in subsequent chapters: the *coupling* technique.[3]

For probability distributions π and σ over Ω, we define

$$d_{TV}(\pi, \sigma) = \max_{X \subseteq \Omega} \left\{ \sum_{x \in X} \pi(x) - \sigma(x) \right\} \tag{1.1}$$

interpreting the summation as an integral if that is appropriate for Ω.

Note that

$$d_{TV}(\pi, \sigma) = \sum_{x \in \Omega : \pi(x) > \sigma(x)} \pi(x) - \sigma(x)$$

$$= \sum_{x \in \Omega : \sigma(x) > \pi(x)} \sigma(x) - \pi(x)$$

$$= \frac{1}{2} \sum_{x \in \Omega} |\pi(x) - \sigma(x)|.$$

Variation distance is equivalent to one half of the ℓ_1 norm. It is also referred to by some authors as *stochastic distance*.

Since variation distance takes a value between zero and one, it has an interpretation as a probability: it is the probability of *error* in the sampling, i.e. if we wish to sample from σ, but can only sample from π, then we may regard π as an approximate sampler for σ that samples with the correct distribution with probability $1 - d_{TV}(\pi, \sigma)$, and samples with some other (incorrect) distribution with probability $d_{TV}(\pi, \sigma)$.

PROPOSITION 1. *If X and Y are random variables taking values over the range $[a, b]$, and $d_{TV}(X, Y) \le \varepsilon$, then $\mathbf{E}(Y) - (b - a)\varepsilon \le \mathbf{E}(X) \le \mathbf{E}(Y) + (b - a)\varepsilon$.*

Proof. The proof is immediate from the definition of variation distance.   □

1.3 Markov Chains

A discrete time random process X may be defined as a collection of random variables, $X_0, X_1, X_2, \ldots$. Such a random process is called a Markov process if for all states a,

---

[3]Some previous work in this area has used a different distance measure, known as *relative pointwise distance* (see, for example, Sinclair [111]), which measures the largest relative difference in probability at any state; also the *chi-squared* distance has been used by, for example, Frieze, Kannan, and Polson [50].

we have

$$\mathbf{P}(X_{i+1} = a \mid X_i) = \mathbf{P}(X_{i+1} = a \mid X_0, X_1, \ldots, X_i).$$

A Markov chain, or random walk, is a Markov process on a finite or countable state space.

If, for all times i, and all pairs of states a and b, we have that

$$\mathbf{P}(X_{i+1} = a \mid X_i = b) = \mathbf{P}(X_1 = a \mid X_0 = b),$$

then the Markov process is said to be *homogeneous*.

We may interpret a homogeneous Markov chain X in other ways: for example, we may interpret it as an initial distribution on states, together with a transition matrix $P = p_{a,b}$, where

$$p_{a,b} = \mathbf{P}(X_1 = b \mid X_0 = a).$$

Alternatively we may interpret the transition matrix as a weighted directed graph, with vertices being the set of states, and a directed edge from a to b taking weight $p_{a,b}$, the sum of the out-edges from each vertex being 1. This latter condition is equivalent to the transition matrix having column sums equal to 1.

Clearly each of these representations of a homogeneous Markov chain leads to an equally valid definition, and we move freely between these isomorphic representations.

A homogeneous Markov chain is said to be *irreducible* if and only if there is a directed path between each pair of states that comprises only directed edges with positive weights.

A Markov chain is said to be *aperiodic* if for each state a and time i, the set of times $\{n > i : \mathbf{P}(X_n = a \mid X_i = a) > 0\}$ has greatest common divisor 1. A Markov chain is said to be *strongly aperiodic* if for each state a and time i, we have $\mathbf{P}(X_{i+1} = a \mid X_i = a) > \frac{1}{2}$.

A stationary distribution for a homogeneous Markov chain is a distribution on the states with the property that if X_0 "is" the stationary distribution, then so is X_1, and hence X_i for all i.

A homogeneous Markov chain that is irreducible and aperiodic has a unique stationary distribution [58]. Such a Markov chain is said to be *ergodic*.

There is a standard theorem (see e.g. [58, Theorem 6.5.3]) which states: for an irreducible Markov chain, if there exists a probability distribution, π, of its states, satisfying the (detailed) balance equation

$$\pi(\alpha)\mathbf{P}(X_{t+1} = \beta \mid X_t = \alpha) = \pi(\beta)\mathbf{P}(X_{t+1} = \alpha \mid X_t = \beta) \tag{1.2}$$

then π is a stationary distribution of the Markov chain. If the detailed balance condition is satisfied, then the Markov chain is said to be *time reversible*. Equivalently, if we define D to be the diagonal matrix with entries $D_{ii} = \pi(i)$, then the Markov chain is time reversible if and only if $D^{1/2}PD^{-1/2}$ is symmetric.

If we restrict ourselves to finite state spaces, it is worth noting that the transition matrix is amenable to treatment by linear algebra. In particular, we may consider the eigenvalues of this matrix. The eigenvalues of an ergodic Markov chain on a finite state space (of size N) are conventionally written as $1 = \lambda_1 \geq \lambda_2 \geq \cdots \geq \lambda_N > -1$.

At this stage, we should introduce some more terminology. The *mixing rate* of an ergodic Markov chain, which will be a function of ε, and typically, the problem size, is the time taken for the Markov chain to reach to within a variation distance of ε of stationarity, maximized over all initial distributions. We will write the mixing rate as $\tau(\varepsilon) = \tau_{\mathcal{M}}(\varepsilon)$. The *mixing time* is the time for the Markov chain to reach within a variation distance of $1/e$ of stationarity, i.e., equal to $\tau(1/e)$. We shall abbreviate this to τ where this will not cause confusion.

Define the *effective mixing rate* of a Markov chain, $\tau_x(\varepsilon)$ to be the time for the Markov chain to reach within a variation distance of ε of stationarity, given initial state x.

Remark. Note that different authors use slightly different terminology. Aldous [1], uses the term mixing time to mean $\tau(1/2e)$. Aldous and Fill [3] also use $1/2e$, but call the time "the variation threshold time". This use of different constants is unimportant, as we will see in the following proposition. $\bigcirc$

PROPOSITION 2. *For any constant $c \in (0,1)$, and any time homogeneous Markov chain, $\Theta(\tau(c)) = \Theta(\tau)$.*

We shall defer the proof of this proposition until Section 2.3.1.1.

Another term that we shall mention is the *relaxation time*, written τ_2. Informally, it is essentially the mixing time, but averaged instead of maximized over all possible starting states. Formally, it is defined in terms of the second-largest eigenvalue of the transition matrix of the Markov chain. Specifically, if λ_2 is the second-largest eigenvalue, then we define

$$\tau_2 = \frac{1}{1 - \lambda_2}.$$

The following lemma draws heavily on some work of Dyer and Greenhill [43], although the result is not made explicit there.

LEMMA 1. *Suppose $\mathcal{M}$ is an irreducible Markov chain with stationary distribution π, and finite state space Ω. Let ν_t be the distribution of $\mathcal{M}$ after t steps, when the initial state is drawn from a distribution ν_0.*

If, for all distributions ν_0, some infinite strictly increasing sequence of natural numbers, $i_1, i_2, \ldots$, and some constant, A, the Markov chain $\mathcal{M}$ is such that

$$d_{\mathrm{TV}}(\nu_{i_t}, \pi) \leq Ac^{i_t},$$

then the relaxation time of $\mathcal{M}$ is bounded above by $1/(1 - c)$.

Proof. Let P be the transition matrix of $\mathcal{M}$. Let $\sigma = \min\{\pi(x) : x \in \Omega, \pi(x) > 0\}$. We will use $\|\cdot\|$ to denote the ℓ_2 norm. Let v be a left eigenvector of P, with eigenvalue λ_2, such that $\|v\| = 1$. Define $\mu = \pi - \sigma v$, and e to be the vector of 1s. Note that every component of μ is non-negative, and that e is a right eigenvector of P (with unit eigenvalue).

Now $v.e = vPe^{\mathsf{T}} = \lambda_2 v.e$, since P is a stochastic matrix, and v is one of its eigenvectors. Since $\lambda_2 \neq 1$, it follows that $v.e = 0$. Thus $\mu.e = (\pi - \sigma v).e = \pi.e = 1$.

Since μ is non-negative, and $\mu.e = 1$, we see that μ is a probability distribution, and hence $\mu.P^t$ is a probability distribution too.

Thus

$$
\begin{aligned}
Ac^{i_t} &\geq \mathrm{d_{TV}}(\mu P^{i_t}, \pi) & \text{(by hypothesis)} \\
&\geq \tfrac{1}{2}\|\mu P^{i_t} - \pi\| & \text{(by the Cauchy–Schwarz inequality)} \\
&= \tfrac{1}{2}\|(\mu - \pi)P^{i_t}\| & \text{(since } \pi \text{ is the stationary distribution)} \\
&= \sigma/2\,\lambda_2^{i_t}\|v\| = \sigma/2\,\lambda_2^{i_t} & \text{(by the definitions of } \sigma \text{ and } v).
\end{aligned}
$$

Thus

$$
\log \lambda_2 \leq \frac{\log {}^{2A}/_{\sigma}}{i_t} + \log c,
$$

and taking limits as $t \to \infty$, we see that $\lambda_2 \leq c$ from which the result is immediate. $\qquad\square$

Sinclair [110, Proposition 1] provides us with another result which links effective mixing rates and relaxation time:

PROPOSITION 3. *For an ergodic time-reversible Markov chain on a finite state space, with stationary distribution π,*

$$
\tau_x(\varepsilon) \leq \tau_2(\ln \pi(x)^{-1} + \ln \varepsilon^{-1}).
$$

1.4   Graph Theory

The fundamentals of graph theory may be gleaned from any undergraduate text, a readable guide being Wilson [130]. We will very briefly go through some definitions for completeness.

A *graph* $G = (V, E)$, consists of a set V of *vertices* (or *nodes*), and E *edges*. We will typically take E to be a set of ordered pairs of vertices. In an *undirected graph*, we will insist that for $a, b \in V$, either both or neither (a, b) and (b, a) is in E. The edge set of a *directed graph* has no such restriction. An edge (a, a) is called a *loop*. On occasions we will deal with *multigraphs* in which there may be multiple edges between the same pair of vertices.

The *degree* of a vertex is simply a count of the number of edges incident upon the vertex. For a directed graph, the *in-degree* and *out-degree* of a vertex are defined respectively to be the number of edges oriented away from and towards the vertex in question.

The set of edges of a graph may be considered to be a relation on the vertices. Consider the equivalence relation given by the reflexive transitive closure of the edge set. The number of *components* of a graph is the number of equivalence classes in this relation. A graph with only one such equivalence class is said to be *connected*.

A graph has a *cycle* if it is possible to traverse a distinct sequence of edges from vertex to vertex, so as to end up at the original vertex.

A *forest* is a graph with no cycles. A *tree* is a connected forest. A *complete graph* is a graph with an edge between every pair of vertices.

An *independent set* of a graph is a subset of the vertices of the graph that does not contain both end-points of any edge.

An *orientation* of a graph is a directed graph, such that for each edge (a, b) in the original graph, exactly one of (a, b) or (b, a) is an edge of the directed graph.

If we *delete* an edge e of a graph $G = (V, E)$, we form a new graph $G - e$, which has the same vertex set, and edge set $E \setminus e$.

If we *contract* an edge $e = (a, b)$ of a graph $G = (V, E)$, we form a new graph $G \cdot e$, which has vertex set $V \setminus b$ (we use the symbol $\setminus$ to denote set minus), and an edge set that is formed from E by replacing each edge incident on b with one incident on a, i.e. (c, b) becomes (c, a) for all c.

A *proper graph colouring* is a mapping from the set of vertices of a graph to a set of colours, such that the end-points of each edge are mapped to different colours.

A *matching* of a graph is a subset of the edges such that no two edges share a vertex. A *perfect matching* is a matching that covers all of the vertices in the graph.

A *bipartite*, or more generally k-*partite*, graph is a graph in which the vertices may be partitioned into 2 (or generally k) subsets, such that no edge has end-points in two partitions.

1.4.1   Tutte–Gröthendieck Polynomial

The Tutte–Gröthendieck polynomial, or simply Tutte polynomial, $T(G; x, y)$ of a graph G contains a remarkable wealth of combinatorial information about G. For example, its evaluation at different values of x and y gives:

- the number of spanning trees of G,

- the number of forests of G,

- the number of acyclic orientations of G, and

- the number of k-colourings of G, for all k.

amongst many others. For a more comprehensive list, see e.g. [6, 126] and the references therein.

The definition of the Tutte polynomial can be phrased remarkably simply in terms of a recursive formula (see, e.g. [6, 126][4]). If G has no edges, then $T(G; x, y) \equiv 1$, otherwise, for *any* edge e in G:

1. If e is a *loop*, i.e. an edge with both end-points the same, then $T(G; x, y) \equiv yT(G - e; x, y)$.

2. If e is an *isthmus*, i.e. an edge, the deletion of which would separate its end points into two components, then $T(G; x, y) \equiv xT(G \cdot e; x, y)$.

---

[4]In fact, [6] *misdefines* the Tutte polynomial, but what is actually meant is clear.

3. If e is neither a loop nor an isthmus, then $T(G; x, y) \equiv T(G \cdot e; x, y) +$ T(G - e; x, y).

We comment that this recursive definition does lead to a unique definition of T.

We will encounter the Tutte polynomial in Sections 4.2.3, 5.3.3, and 5.4.

1.4.2 Hypergraphs

A hypergraph is the generalization of a graph, such that edges can *cover* (or *contain*) an arbitrary number of vertices.

Formally, a hypergraph G consists of a set of vertices, and a set of edges, each of which is a subset of the set of vertices. Some concepts transfer naturally from the equivalent graph definition, e.g. two vertices in a hypergraph are said to be *adjacent* if there is an edge in which they are both contained. We will write $v \sim w$ if $v \neq w$ and $\{v, w\} \subseteq e$, for some $e \in E$, and $N(v) = \{w : v \sim w\}$. A hypergraph is said to be *connected* if the reflexive transitive closure of the adjacency relation is the complete relation.

The concept of the *degree* of a vertex, and hence of the hypergraph may be extended from that of a graph in several ways. In this book we will make use of four degree functions.

$$\delta_1(v) = \max_{S \subseteq E} \{|S| : \forall a, b \in S \bullet a \neq b \Rightarrow a \cap b = \{v\}\},$$

$$\delta_2(v) = |\{e \in E : v \in e\}|,$$

$$\delta_3(v) = |N(v)|, \text{ and}$$

$$\delta_4(v) = \max_{S \subseteq N(v)} \{|S| : \forall a, b \in S, e \in E, \{a, b, v\} \not\subseteq e\}.$$

A hypergraph G is said to have degree $\Delta_i = \max_{v \in V} \delta_i(v)$ ($i \in \{1, 2, 3, 4\}$). All of these definitions of degree are in accord with those for (simple) graphs. Note that for every vertex, v, $\delta_1(v) \leq \delta_4(v) \leq \min\{\delta_2(v), \delta_3(v)\}$, and hence $\Delta_1 \leq \Delta_4 \leq \min\{\Delta_2, \Delta_3\}$.

The definition of the degree function, δ_1, is standard, and may be found, for example, in Berge [15] or Tomescu [120].

We also define the *co-degree* of two distinct vertices, by $\delta(i, j) = |\{e \in E : \{i, j\} \subseteq e\}|$.

Chapter 2

Techniques for Sampling and Approximate Sampling

In this chapter, we review contemporary techniques for sampling from a set of combinatorial objects. We initially consider some simple methods for sampling exactly from a desired distribution, before turning to the techniques which form the basis for much of the work in this book: approximate sampling via the Markov chain method.

2.1   Introduction

Frequently, when dealing with a set of combinatorial objects, one wishes to determine some information about the set as a whole. For large sets, it may not be practical to examine every element; instead one attempts to form some impression of the set as a whole by considering some statistical properties of a randomly chosen element of the set.

Random samples are also of great practical use as subroutines in other randomized algorithms: a trivial example would be to pick a random permutation to transform the input to a sorting algorithm, say.

Another significant use of sampling is in approximate counting, a problem which we go into in detail in the next chapter. In essence, when given a set of objects satisfying some property, one may not know just how many objects are in the set, and counting them by enumeration may be too time-consuming. Instead it may be sufficient to approximate their number, especially if this is substantially quicker. It transpires that if we can sample approximately, then in a large class of structures, we can also approximately count (see Section 3.3).

Sampling is by no means a trivial problem for which to provide a fast algorithm unless the set we wish to sample from has a very simple structure. In this book, we consider a set *hard to sample from* (or simply a *hard set*), if the associated counting problem is #P-hard. In the next section we consider a variety of potentially fast methods for sampling — some of these have applications even when the set is hard to sample from.

2.1.1 Definitions

An *almost uniform sampler* of a set Ω is a probabilistic algorithm which takes as input (a suitable encoding of) Ω, and an error parameter, ε, and outputs a random element of Ω: the variation distance of the output distribution of the algorithm, and the uniform distribution on Ω being at most ε. For a detailed discussion of a suitable formal model of computation, see Jerrum, Valiant, and Vazirani [71], or Sinclair [111].

The sampler is said to be a *fully polynomial* almost uniform sampler, if it runs in time that is bounded by a polynomial in the size of the encoding (which will typically be logarithmic in $|\Omega|$) and $\log \varepsilon^{-1}$.

If the distribution that we are seeking to sample from is not uniform, then we will call such an algorithm an *approximate sampling* algorithm.

2.2 Direct Sampling

If there is some known polynomial-time computable bijection from the set of integers $\{1, 2, \ldots, n\}$ to the set from which we wish to sample, then there is an obvious sampling procedure. In the following sections we consider more powerful techniques that may be used when no such bijection has been found.

2.2.1 Monte Carlo Method: Rejection Sampling

Possibly the simplest method for sampling from a complex combinatorial structure is by using a Monte Carlo method, often known as rejection sampling.

Suppose we can embed our target set within another, simpler combinatorial structure, that we can sample from. Then we repeatedly sample from this simpler set until we obtain a sample within our target set.

Although a very simple procedure, and one that will almost surely terminate eventually, one must be careful to prove that not too many samples will be rejected before we obtain a sample in our target set.

Examples of such results may be found in Alon, Frieze and Welsh [6], where it is proved (among other results) that one can use such a rejection sampling scheme to sample from the set of strong orientations of a sufficiently dense graph, by generating random orientations and rejecting those that are not strong.

2.2.2 Karp–Luby Technique

There is another technique, that is in some sense a generalization of the rejection sampling technique of the previous section. Its description, however, lies more naturally in the context of approximate counting rather than of sampling, and accordingly we will defer its further discussion until Chapter 3.

2.3 Markov Chain Method

In cases in which direct sampling techniques cannot be applied, we must seek other methods. So far, only one other method has been developed for sampling: the Markov chain method. Here, we take advantage of the fact that often an approximate sample is all that is needed (but see Section 2.3.7 for how these ideas may be applied to get exact samples too).

Suppose we wish to approximately sample from a distribution π. One first needs to construct a Markov chain with stationary distribution π, and then simulate this Markov chain long enough for the distribution to approximate that of π sufficiently closely.

The crux of the analysis is of course, determining how long "long enough" is for a particular Markov chain. The remainder of this section is devoted to different methods for answering this question. One of these methods, coupling, forms the basis for much of the new research in this book. We consider a variety of applications of this method in Chapter 4. We go on to develop a method for simplifying the analysis of coupling in the Intermezzo (page 83): this we name path coupling. We give some applications of path coupling in Chapter 5.

When using the Markov chain method, in order to ensure that we have a fully-polynomial sampler, we need only to ensure that we can simulate enough steps of the Markov chain in time polynomial in the problem size and in $\log 1/\varepsilon$. Note that there is an implicit requirement here that each step of the Markov chain requires only time polynomial in the problem size to simulate. It is sufficient then to prove that the Markov chain converges to within a variation distance of ε of stationarity in time polynomial in the problem size and in $\log 1/\varepsilon$. Note that it may not be necessary for the Markov chain itself to converge that quickly, provided that we can *simulate* that number of steps in polynomial time. In Section 4.3.4.1 we see how we can simulate $O(n)$ steps of a particular Markov chain by taking essentially only $O(\log n)$ steps. There is thus no *a priori* reason to suppose that one could not simulate an exponential number of steps of some Markov chain in polynomial time.

The term *rapid-mixing*, when applied to Markov chains was coined by Aldous [1], in the context of random walks on groups; he defined it loosely as "the property that τ is small compared to $\#G$ [the size of the state space]". Its meaning has subsequently firmed up through usage, and its conventional meaning now is that the Markov chain has a mixing time that is polynomial in the size of the encoding of the problem (which is usually logarithmic in the size of the state space). When we use this term it will always have this meaning.

If a Markov chain is rapidly mixing, and may be simulated quickly, then the Markov chain method will be a fast algorithm for approximate sampling.

Frequently, in applications, our motive in drawing approximate samples is to enable us to estimate the average of some function on the state space; probably the most significant example is when we are drawing approximate samples as part of an approximate counting algorithm.

A naïve method for estimating this average, with error of order ε, would be to run $\Theta(\varepsilon^{-2})$ consecutive simulations, each for time $\Theta(\tau)$, and take the resulting sample average.

A better method is one proposed by Aldous [2], although a clearer exposition may be found in Aldous and Fill [3]. Essentially, this method simulates the Markov chain for $\Theta(\tau) + \Theta(\varepsilon^{-2}\tau_2)$ steps, averaging the function over all steps after the initial $\Theta(\tau)$ steps (in the statistical physics community, this is referred to as the initial "burn-in" time).

In the following sections, we consider a variety of proof techniques for proving rapid mixing, and, in particular, for bounding mixing and relaxation times.

It should be noted that in cases of interest, the state space of the Markov chain will be super-polynomial in the size of the problem, and thus recent advances (see e.g. Pilarski and Kameda [99]) in calculating convergence rates by analyzing the transition matrix itself cannot yield "fast" algorithms in the context of this book.

2.3.1   Coupling

The coupling argument, in its simplest form, appears as far back as the 1930s in the work of Doeblin.

Central to the coupling technique is the "Coupling Lemma", which in this form is usually attributed to Aldous [1]:

COUPLING LEMMA (LEMMA 2). *Suppose we have a joint process, (X, Y), not necessarily Markovian, such that marginally, X and Y are both Markov chains with the same transition probabilities and state space. If μ_t and ν_t are their respective distributions at time t, then*

$$d_{\mathrm{TV}}(\mu_t, \nu_t) \leq \mathbf{P}\left(X_t \neq Y_t\right).$$

Such a joint process is called a *coupling*.

Hence, a sufficient condition to achieve a variation distance of ε is to simulate the Markov chain for long enough to ensure that $\mathbf{P}\left(X_t \neq Y_t\right) \leq \varepsilon$, and thus $\tau(\varepsilon) \leq \min\{t : \mathbf{P}(X_t \neq Y_t) \leq \varepsilon\}$.

Proof of Coupling Lemma. The proof of the coupling lemma is remarkably simple, and follows almost immediately from the definition of variation distance. For a state space Ω, we have from Equation (1.1):

$$d_{\mathrm{TV}}(\mu_t, \nu_t) = \max_{X \subseteq \Omega}\left\{\sum_{x \in X} \mu_t(x) - \nu_t(x)\right\}.$$

If we define $\mu_{t,=}$ and $\nu_{t,=}$ to be the distributions of X_t and Y_t conditioned on the fact that $X_t = Y_t$, and *mutatis mutandis* define $\mu_{t,\neq}$ and $\nu_{t,\neq}$, then we may write:

$$
\begin{aligned}
d_{\text{TV}}&(\mu_t, \nu_t) \\
&= \max_{X \subseteq \Omega} \left\{ \mathbf{P}(X_t = Y_t) \sum_{x \in X} (\mu_{t,=}(x) - \nu_{t,=}(x)) + \right. \\
&\qquad\qquad\qquad \left. \mathbf{P}(X_t \neq Y_t) \sum_{x \in X} (\mu_{t,\neq}(x) - \nu_{t,\neq}(x)) \right\} \\
&= \mathbf{P}(X_t \neq Y_t) \max_{X \subseteq \Omega} \left\{ \sum_{x \in X} \mu_{t,\neq}(x) - \nu_{t,\neq}(x) \right\} \\
&\leq \mathbf{P}(X_t \neq Y_t).
\end{aligned}
$$

$\square$

Coupling has been employed in a number of similar contexts to those in this book.

- Broder [18] made the first attempt to use the coupling method to prove rapid mixing of a Markov chain in the context of hard sampling problems, when he considered a Markov chain on the set of perfect matchings of a bipartite graph. Unfortunately, the proof was flawed [94], although the result was later shown to be correct (under slightly stronger conditions) by Jerrum and Sinclair [68]. This latter analysis, however, was not through the technique of coupling, but instead used canonical paths (see Section 2.3.3).

- Matthews [92] showed rapid mixing of a geometric Markov chain for linear extensions of a partial order (see Section 5.7).

- Jerrum [67] showed rapid mixing of the Metropolis Markov chain on k-colourings of graphs, for $k \geq 2\Delta + 1$ (see also Sections 4.1, 5.8, and 5.10).

- Luby, Randall, and Sinclair [89] used the coupling technique to show rapid mixing of Markov chains on domino tilings (dimer systems), lozenge tilings, and Eulerian orientations of a graph (the six-point ice model).

- Luby and Vigoda [91] showed how to uniformly sample independent sets of a low-degree graph, and, more generally, to sample weighted independent sets in a fashion that corresponds to the Gibbs distribution of the hard-core model in statistical physics. This problem has seen several follow-up papers (see Section 2.3.5).

In addition to the above results that use the coupling technique, the results of Chapter 4 also use this technique. Some of the results of Chapter 4 have appeared elsewhere as Bubley and Dyer [20], and Bubley, Dyer, and Jerrum [26].

2.3.1.1 Maximal couplings

Couplings form the basis of a very useful theorem, on the existence of *maximal couplings* (see e.g. [12, 118]).

THEOREM 1. *Suppose X and Y are both Markov chains with the same transition probabilities and state space, and with respective distributions μ_t and ν_t at time t. Then there is a joint process (X, Y), such that*

$$\mathbf{P}\left(X_t \neq Y_t\right) = \mathrm{d}_{\mathrm{TV}}(\mu_t, \nu_t).$$

This is a very simple, yet powerful theorem. In some sense it justifies the continued study of coupling techniques for proving bounds on convergence rates, since there must, by the above theorem, be a coupling that enables one to calculate the convergence rates exactly! Whilst we have not managed to push the coupling technique *quite* this far in approximately sampling from hard sets, we get close in Section 5.7, where we illustrate a coupling on the set of linear extensions that is tight to within a constant factor; unfortunately, this is currently the only example of such a tight analysis for a complex Markov chain.

A simple use of Theorem 1 is to prove Proposition 2: for any constant $c \in (0, 1)$, and any Markov chain, $\Theta(\tau(c)) = \Theta(\tau)$.

Proof of Proposition 2. This proof is based on a device of Jerrum [67]. Suppose $c \leq 1/e$, and thus $\tau(c) \geq \tau$. Consider two copies of the Markov chain, X, and Y. By Theorem 1, there is a coupling of X and Y, such that after τ steps, $\mathbf{P}(X_t \neq Y_t) \leq 1/e$. Suppose we run successive independent coupling *trials*, each of length τ, duplicating the coupling procedure within each trial. We have thus constructed a coupling such that after $k\tau$ steps, the probability that $X_t \neq Y_t$ is bounded above by $1/e^k$. If we take $k = \lceil -\ln c \rceil$, we have immediately that $\tau(c) \leq k\tau$, and thus $\Theta(\tau(c)) = \Theta(\tau)$.

The case where $c > 1/e$ is essentially the same. $\square$

2.3.2 Dobrushin's Uniqueness Criterion

This proof technique is one that is little known outside the statistical physics community. Within the statistical physics literature, the technique is used to prove the existence of a unique stationary distribution to certain Markov chains on infinite state spaces, whence the eponymous uniqueness comes. In particular a close relative of Lemma 3 may be found in a variety of works [53, 59, 83, 109, 113]. The results presented here are more general than those found in the statistical physics literature. An earlier version of this result may be found in work of Bubley and Dyer [27], and there are close connexions between this idea and the concept of path coupling, which we introduce in the Intermezzo (page 83). Peinado and Lengauer [98] have recently introduced a different generalization, motivated by a problem from computational chemistry.

We give a generalization of what is usually proved in the literature (although, of course, our concerns here are somewhat different). The criterion is normally given for the so-called "heat bath dynamics", a particular form of Markov chain. Our proof is entirely general and does not even require time reversibility.

Although we prove results only for a particular class of Markov chains, this is not to say that the method cannot be generalized further. The method is very similar in some ways to the path coupling method, which we introduce in the Intermezzo, and indeed it was the study of Dobrushin's uniqueness criterion that led us to the path coupling idea.

On the downside, we are currently unaware of any direct applications of Dobrushin's uniqueness criterion that cannot also be achieved or bettered by the path coupling method. Nevertheless, we give an application of Dobrushin's uniqueness criterion in Appendix A, as a "proof-of-concept".

Dobrushin's uniqueness criterion has also been used recently by Salas and Sokal [106] to prove some results on graph colouring and the Potts model. These results may be recaptured easily by path coupling. They go on, however, to provide an indirect computational application of Dobrushin's uniqueness criterion (called "decimation"). This is applied to the problem of colouring some lattice graphs; it is not clear if this result may be recaptured by path coupling too.

Let V and C be finite sets, with $n = |V|$ and $k = |C|$. We consider a finite Markov chain $\mathcal{M}$, with state space $\Omega = C^V$ and unique equilibrium distribution π.

For $X \in \Omega$, $v \in V$, and $c \in C$ let us use the notation $X_{v \to c}$ to denote the state resulting from making the transition at X associated with the pair (v, c). Thus

$$X_{v \to c}(w) = \begin{cases} c & \text{if } w = v, \text{ and} \\ X(w) & \text{otherwise.} \end{cases}$$

Using this notation, we may more precisely define the transition structure of $\mathcal{M}$. We first pick $v \in V$ from the uniform distribution on V. Then we pick $c \in C$ according to a distribution $\nu_{X,v}$ on C, dependent only on the current state X and v, and make the transition to $X_{v \to c}$. We assume that $X_{v \to c} \notin \Omega$ implies that $\nu_{X,v}(c) = 0$.

THEOREM 2. *Let $\Omega = C^V$, and let J be the uniform distribution. Define*

$$\varrho_{ij} = \max_{X,Y \in \Omega} \left\{ d_{TV}\left(\nu_{X,j}, \nu_{Y,j}\right) \mid Y = X_{i \to c} \text{ for some } c \in C \right\},$$

$$\alpha = \max_{j \in V} \left\{ \sum_{i \in V} \varrho_{ij} \right\}.$$

Then provided $\alpha < 1$, $\tau(\varepsilon) \le \lceil n \ln \left(n \varepsilon^{-1} \right) / (1 - \alpha) \rceil$.

Before we start the proof, we will make some further definitions. Assume that $|\Omega| = q$, and that Ω is enumerated by $\omega_1, \omega_2, \ldots, \omega_q$. Then, for a function $f : \Omega \to \mathbb{R}$, define linear maps $P^{(v)}$ and P by:

$$\left(P^{(v)} f\right)(X) = \sum_{c \in C} \nu_{X,v}(c) f(X_{v \to c}), \quad \text{and} \quad Pf = \frac{1}{n} \sum_{v \in V} P^{(v)} f.$$

Observe that if $\tilde{f} = \left(f\left(\omega_1\right) f\left(\omega_2\right) \ldots f\left(\omega_q\right)\right)^{\top}$, then we could equivalently have defined the linear maps $P^{(v)}$ and P in terms of matrices $\tilde{P}^{(v)}$ and $\tilde{P}$, so that $\left(\tilde{P}^{(v)} \tilde{f}\right)_i$

$= \left(P^{(v)}f\right)(\omega_i)$, (with $\tilde{P}$ defined in the same way), where $(a)_i$ denotes the ith element of column vector a. Then $\tilde{P}$ is the transition matrix of the Markov chain $\mathcal{M}$.

As an interpretation, $\left(P^{(v)}f\right)(X)$ is the expectation of $f(X)$, after X has undergone a transition at v. Furthermore, since v is chosen uniformly at random in a transition, $(Pf)(X)$ is the expectation of $f(X)$ after X has undergone a transition (unconditioned on v).

Finally define

$$\gamma_i(f) = \max_{X,Y \in \Omega} \left\{ |f(X) - f(Y)| \mid Y = X_{i \to c} \text{ for some } c \in C \right\}, \quad \text{and}$$

$$\Gamma(f) = \sum_{j \in V} \gamma_j(f).$$

Observe that the γ_i, and hence Γ are semi-norms, i.e. for reals $\lambda, \mu \geq 0$ and functions f, g, we have $\Gamma(\lambda f + \mu g) \leq \lambda\Gamma(f) + \mu\Gamma(g)$.

LEMMA 3. $\gamma_i(P^{(j)}f) \leq \epsilon_{ij}\gamma_i(f) + \varrho_{ij}\gamma_j(f)$, *for all* $i, j \in V$, *where* ϵ *is the complement of the Kronecker delta, i.e. if* $i \neq j$, $\epsilon_{ij} = 1$, *but for all* i, $\epsilon_{ii} = 0$.

Proof of Lemma 3. Suppose X and Y are in Ω, and that $Y = X_{i \to c}$ for some $c \in C$. For a function $h : C \to \mathbb{R}$, we will use $h^+(\cdot)$ to denote $\max_{c \in C}\{h(c)\}$, and $h^-(\cdot)$ to denote $\min_{c \in C}\{h(c)\}$. Then:

$$\left| \left(P^{(j)}f\right)(X) - \left(P^{(j)}f\right)(Y) \right|$$

$$= \left| \sum_{c \in C} f(X_{j \to c})\nu_{X,j}(c) - \sum_{c \in C} f(Y_{j \to c})\nu_{Y,j}(c) \right|$$

$$\leq \sum_{c \in C} |f(X_{j \to c}) - f(Y_{j \to c})| \, \nu_{X,j}(c) + \left| \sum_{c \in C} f(Y_{j \to c}) [\nu_{X,j}(c) - \nu_{Y,j}(c)] \right|$$

$$\leq |f(X_{j \to \cdot}) - f(Y_{j \to \cdot})|^+$$

$$+ \left| \sum_{c \in C^+} f(Y_{j \to c}) [\nu_{X,j}(c) - \nu_{Y,j}(c)] + \sum_{c \in C^-} f(Y_{j \to c}) [\nu_{X,j}(c) - \nu_{Y,j}(c)] \right|,$$

where $C^+ = \{c \in C : \nu_{X,j}(c) \geq \nu_{Y,j}(c)\}$,

and $C^- = \{c \in C : \nu_{X,j}(c) < \nu_{Y,j}(c)\}$,

$$\leq \epsilon_{ij}\gamma_i(f)$$

$$+ \left| f(Y_{j \to \cdot})^+ \sum_{c \in C^+} [\nu_{X,j}(c) - \nu_{Y,j}(c)] + f(Y_{j \to \cdot})^- \sum_{c \in C^-} [\nu_{X,j}(c) - \nu_{Y,j}(c)] \right|$$

$$\leq \epsilon_{ij}\gamma_i(f) + \left| f(Y_{j \to \cdot})^+ \varrho_{ij} - f(Y_{j \to \cdot})^- \varrho_{ij} \right|$$

$$= \epsilon_{ij}\gamma_i(f) + \varrho_{ij}\gamma_j(f).$$

Since this inequality holds for arbitrary $X, Y \in \Omega$, where $Y = X_{i \to c}$, the proof is established. $\qquad\square$

COROLLARY 1.

$$\Gamma(Pf) \le \left(1 - \frac{1-\alpha}{n}\right)\Gamma(f).$$

Proof. Taking the inequality provided by Lemma 3 above, and summing over all $i \in V$, we obtain

$$\Gamma(P^{(j)}f) \le \sum_{i \in V} \epsilon_{ij}\gamma_i(f) + \varrho_{ij}\gamma_j(f).$$

Summing now over all $j \in V$, and recalling that Γ is a semi-norm, we have

$$\begin{aligned}
\Gamma(Pf) = \Gamma\left(n^{-1}\sum_{j \in V} P^{(j)}f\right) &\le n^{-1}\sum_{j \in V}\Gamma(P^{(j)}f) \\
&\le n^{-1}\sum_{i,j \in V}[\epsilon_{ij}\gamma_i(f) + \varrho_{ij}\gamma_j(f)] \\
&= \frac{n-1}{n}\Gamma(f) + \frac{1}{n}\sum_{j \in V}\left(\sum_{i \in V}\varrho_{ij}\right)\gamma_j(f).
\end{aligned}$$

Recall the definition $\alpha = \max_{j \in V}\left\{\sum_{i \in V}\varrho_{ij}\right\}$. Thus

$$\Gamma(Pf) \le \frac{n-1}{n}\Gamma(f) + \frac{1}{n}\sum_{j \in V}\alpha\gamma_j(f),$$

from which the result follows. $\square$

LEMMA 4. *Let* $M(f) = \max_{X \in \Omega}\{f(X)\} - \min_{X \in \Omega}\{f(X)\}$. *Then* $M(f) \le \Gamma(f) \le nM(f)$.

Proof. Let X' and X'' be such that

$$f(X') = \max_{X \in \Omega}\{f(X)\}, \text{ and,} f(X'') = \min_{X \in \Omega}\{f(X)\}.$$

Let $h = H(X', X'')$, the Hamming distance between X' and X'', and let $X'' = Z_0, Z_1, \ldots, Z_h = X'$ be a sequence such that $Z_i = (Z_{i-1})_{v_i \to c_i}$ for some $v_i \in V, c_i \in C$ $(i = 1, 2, \ldots, h)$. Then

$$M(f) = f(X') - f(X'') = \sum_{i=1}^{h}[f(Z_i) - f(Z_{i-1})] \le \sum_{i \in V}\gamma_i(f) \le \Gamma(f).$$

For the second inequality, it is enough to note that for all i, $\gamma_i(f) \le M(f)$, and thus $\Gamma(f) \le nM(f)$. $\square$

We are now in a position to prove Theorem 2.

Proof of Theorem 2. We apply Lemmas 3 and 4. Suppose σ is the initial distribution of $\mathcal{M}$. Let $\tilde{\sigma}$ denote the vector $(\sigma(\omega_1)\,\sigma(\omega_2)\,\ldots\,\sigma(\omega_q))$, and similarly $\tilde{\pi}$ the vector $(\pi(\omega_1)\,\pi(\omega_2)\,\ldots\,\pi(\omega_q))$. Then $\tilde{\sigma}\tilde{P}^r$ is the distribution after r steps.

So, by the definition of total variation distance (and a slight abuse of notation), $d_{TV}(\tilde{\pi}, \tilde{\sigma}\tilde{P}^r) = \frac{1}{2}\sum_{i=1}^{q}|(\tilde{\pi})_i - (\tilde{\sigma}\tilde{P}^r)_i| = (\tilde{\pi} - \tilde{\sigma}\tilde{P}^r)\tilde{f}$, where $f(X) \in \{-\frac{1}{2}, \frac{1}{2}\}$, and hence $M(f) \le 1$. Thus, since $\tilde{\pi} = \tilde{\pi}\tilde{P}$,

$$d_{TV}(\tilde{\pi}, \tilde{\sigma}\tilde{P}^r) \le \tilde{\pi}\tilde{P}^r\tilde{f} - \tilde{\sigma}\tilde{P}^r\tilde{f} \le M(P^r f),$$

as $\tilde{\pi}\tilde{P}^r\tilde{f} \le \max_{X\in\Omega}\{P^r f(X)\}$ and $\tilde{\sigma}\tilde{P}^r\tilde{f} \ge \min_{X\in\Omega}\{P^r f(X)\}$. But $M(P^r f) \le \Gamma(P^r f)$ by Lemma 4, and $\Gamma(P^r f) \le (1 - (1 - \alpha)/n)^r \Gamma(f)$ by the Corollary to Lemma 3. This in turn is bounded above by $n(1 - (1 - \alpha)/n)^r M(f)$ by a second application of Lemma 4. However, as we noted earlier, $M(f) \le 1$ in this instance, and so

$$d_{TV}(\pi, \sigma\tilde{P}^r) \le n\left(1 - \frac{1-\alpha}{n}\right)^r \le n\exp\left(-\frac{r(1-\alpha)}{n}\right).$$

Taking logarithms and rearranging this inequality completes the proof.   □

It is interesting to compare this theorem with the following result which we present in the Intermezzo.

COROLLARY OF THEOREM 4 (COROLLARY 2). *Let* $\Omega = C^V$, *J be the uniform distribution and* $\alpha' = \max_{i\in V}\left\{\sum_{j\in V}\varrho_{ij}\right\}$.
Then provided $\alpha' < 1$, $\tau(\varepsilon) \le \lceil n\ln\left(n\varepsilon^{-1}\right)/(1-\alpha')\rceil$.

These are remarkably similar, but involve transposition of the so-called "Dobrushin matrix" (ϱ_{ij}). Thus they are in some sense "dual" results. This duality arises from the Dobrushin computation being carried out in the space of functions $\Omega \to \mathbb{R}$ rather than the state space Ω.

Whilst there appears to be no obvious intuitive reason that this theorem should only apply when $\Omega = C^V$, there appear to be difficulties in generalizing the proof to $\Omega \subseteq C^V$.

The path coupling results deal with two *particular* adjacent states. The sum in Dobrushin's criterion must be evaluated for the worst pair of adjacent states *for each summand*.

2.3.3 Canonical Paths

Historically, the two techniques of canonical paths, and of conductance (see Section 2.3.4) were developed in tandem by Sinclair and Jerrum [112].

The idea behind the canonical paths techniques is to consider a potential route between each pair of states (the canonical paths), and study the resulting *congestion* along such routes.

There are several variations on the canonical paths technique, each of which has slightly different definitions of congestion, and consequently different bounds on the mixing rate, see e.g. Diaconis and Stroock [38] and Sinclair [110]. In order to get a flavour of the type of result in this area, we concentrate on one particular definition of congestion below.

Formally, suppose we have a strongly aperiodic time-reversible Markov chain, $\mathcal{M}$, with state space Ω and distribution ν_t at time t, and define Q from the detailed balance equation:

$$Q(\alpha, \beta) = \pi(\alpha)\mathbf{P}\left(\nu_{t+1} = \beta \mid \nu_t = \alpha\right) = \pi(\beta)\mathbf{P}\left(\nu_{t+1} = \alpha \mid \nu_t = \beta\right).$$

Consider the weighted graph with vertex set Ω, and edge weights being given by Q. Note that each vertex in this graph will have a self-loop.

For each pair of states, $x, y \in \Omega$, define a (canonical) path, γ_{xy} from x to y in this graph, along edges with strictly positive weights. Let $\Gamma = \{\gamma_{xy} : x, y \in \Omega\}$.

Define the congestion, $\bar{\varrho}$, by

$$\bar{\varrho} = \bar{\varrho}(\Gamma) = \max_e \frac{1}{Q(e)} \sum_{\gamma_{xy} \ni e} \pi(x)\pi(y)|\gamma_{xy}|,$$

where the sum is over all directed edges e, such that $Q(e) > 0$, and where $|\gamma_{xy}|$ denotes the length of the path γ_{xy}.

Following Jerrum and Sinclair [66, 70], the following proposition holds:

PROPOSITION 4. *Let $\mathcal{M}$ be a strongly aperiodic time-reversible Markov chain. Let Γ be a set of canonical paths, with congestion $\bar{\varrho}$. Then the effective mixing rate of $\mathcal{M}$ satisfies $\tau_x(\varepsilon) \le \bar{\varrho}(\ln \pi(x)^{-1} + \ln \varepsilon^{-1})$.*

The proof follows from Proposition 1 and Theorem 5 of [110].

A variety of problems have been tackled using the canonical paths technique, some of which are listed below:

- Sampling graphs [112] or bipartite graphs [76] of specified degrees.

- Approximating the permanent [68, 38, 110].

- Sampling all matchings (the monomer–dimer system) [68, 110].

- Sampling from the Gibbs distribution of the ferromagnetic Ising model [69, 110].

Sinclair [110] has also generalized the technique of canonical paths to allow multiple paths (or *flows*) between each pair of states: in some of the examples above this has led to tighter bounds than using simpler canonical paths techniques.

2.3.4 Conductance

The conductance technique was developed alongside canonical paths by Sinclair and Jerrum [112], and independently by Alon [4][1].

The central idea in conductance has a certain intuitive appeal in its simplicity; unfortunately, as is so often the case, the difficulty lies in its application.

---

[1]Cheeger derived a similar result, but in a different context, hence conductance results are sometimes known generically as *Cheeger inequalities*.

Once again, we will suppose we have a strongly aperiodic time-reversible Markov chain, $\mathcal{M}$, with state space Ω and distribution ν_t at time t, and define Q from the detailed balance equation:

$$Q(\alpha, \beta) = \pi(\alpha)\mathbf{P}\left(\nu_{t+1} = \beta \mid \nu_t = \alpha\right) = \pi(\beta)\mathbf{P}\left(\nu_{t+1} = \alpha \mid \nu_t = \beta\right).$$

Then the conductance of $\mathcal{M}$ is defined by:

$$\Phi = \Phi(\mathcal{M}) = \min_{S \subseteq \Omega:\, 0 < \pi(S) \leq 1/2} \frac{Q(S, \bar{S})}{\pi(S)}.$$

If we condition the Markov chain in its stationary distribution to be within a set of states, S, then the conductance is a lower bound on the probability that on its next step the Markov chain will exit from S.

Following Jerrum and Sinclair [66,70], the following proposition holds:

PROPOSITION 5. *Let $\mathcal{M}$ be a strongly aperiodic time-reversible Markov chain. Let Φ be the conductance of $\mathcal{M}$. Then the effective mixing rate of $\mathcal{M}$ satisfies $\tau_x(\varepsilon) \leq 2\Phi^{-2}(\ln \pi(x)^{-1} + \ln \varepsilon^{-1})$.*

The proof follows from Proposition 1 and Theorem 2 of [110].

There have been two main strands of research in deriving bounds on the conductance of a Markov chain. Firstly, the method of canonical paths was originally developed to give a bound on the conductance, and many rapid mixing results thus combined the two techniques. Secondly, a method using isoperimetric inequalities[2] has been developed for use with Markov chains that have a geometric interpretation. We list a few examples of applications of this latter strand below:

- Approximating the volume of a convex body [8,40,42,88].

- Approximately counting linear extensions of a partial order [80].

There have been several generalizations to the definition of conductance proposed in the literature studying the approximation of the volume of a convex body, and the approximate integration of a log-concave function. Lovász and Simonovits [88] define μ-conductance, which allows one to ignore low-probability sets in calculating the conductance (cf. [50]), and, in a somewhat complementary generalization, Kannan [73], and subsequently with co-authors Lovász and Simonovits [74,75], *local conductance*, which considers the conditional exit probability from singleton sets only.

2.3.5   Comparison Methods

Diaconis and Saloff-Coste [35, 36] showed how the relaxation times of two Markov chains on the same state space may be compared, by looking at the differences in their stationary distributions and in their transition matrices. They go on to show how the logarithmic Sobolev constants (see Section 2.3.6.2) may be similarly compared. Randall and Tetali [103] have used these techniques to compare very simple Markov

---

[2]An isoperimetric inequality is an inequality relating the surface area and volume of a body.

chains with more complicated Markov chains on the same state space. They proved rapid mixing of simple Markov chains for sampling weighted independent sets, lozenge and domino tilings, and generating 3-colourings of a subregion of the Cartesian lattice. With the exception of the weighted independent sets it was previously unknown if these simple Markov chains were rapidly mixing. In the case of weighted independent sets, Randall and Tetali built upon earlier results of Luby and Vigoda [90] and Dyer and Greenhill [43].

Dyer and Greenhill [43] used a similar technique to those of Diaconis and Saloff-Coste above, coupled with a result similar to Lemma 1 earlier in this book, to obtain a comparison between the mixing rates of two Markov chains on the same state space with the same stationary distribution.

The primary limitation of these comparison methods is that so far, all of the techniques used result in mixing times for the compared chain that are worse than the mixing times of the chains they are being compared to. Thus, their primary use seems to be in establishing that simple Markov chains are rapid mixing; for algorithmic purposes however, they do not as a consequence give any significant advantage over the Markov chain that they are being compared to.

2.3.6 Other Methods

We saw in Proposition 3 that one could bound the mixing time in terms of the relaxation time and the minimum non-zero value of the stationary distribution. This inequality leaves room for improvement, however, and various techniques have been developed to address this issue. Most of these techniques have their genesis in statistical physics.

2.3.6.1 Poincaré-type inequalities

Frieze, Kannan, and Polson [50], in a paper on sampling from log-concave distributions, introduce the use of what they call a Poincaré-type inequality[3] to analyze the transition matrix of their Markov chain. They observed that in prior solutions to this problem, the analysis of the Markov chain had been ham-strung by a small set of states which have a relatively low probability in the stationary distribution. This led naturally on to a theorem which allowed them to *ignore* these states.

Let λ_2 be the second largest absolute value of an eigenvalue of the transition matrix of our Markov chain.

The starting point was an inequality similar to the ones we saw in Section 1.3[4] i.e.

$$d_{\mathrm{TV}}(\nu_t, \pi) \leq \lambda_2^t \sqrt{\sum_{x \in \Omega} \frac{(\nu_0(x) - \pi(x))^2}{\pi(x)}}$$

---

[3]Diaconis and Stroock [38] also use the term Poincaré inequality, although they use it to refer to the inequalities that we have grouped under canonical paths, see Section 2.3.3.

[4]In fact, the analysis that Frieze, Kannan and Polson gave is in terms of the chi-squared distance between distributions, rather than the variation distance that we have been using in this book; however this has no consequence here.

Suppose S is the set of states we wish to ignore, and $\bar{S} = \Omega \setminus S$ its complement. Define

$$\Phi = \left\{ \phi : \ \phi \in \mathbb{R}^\Omega \setminus \{0\}; \ \pi^\top \phi = 0; \ \text{and } \phi(s) = 0, \ \forall s \in S \right\},$$

$$\lambda = \sup_{\phi \in \Phi} \left\{ \frac{\left| \sum_{x \in \Omega} \phi(x)(P \cdot \phi)(x)\pi(x) \right|}{\sum_{x \in \Omega} \phi^2(x)\pi(x)} \right\}.$$

Then Frieze, Kannan, and Polson prove (among other results) the following bound on variation distance:

$$\mathrm{d_{TV}}(\nu_t, \pi) \leq \lambda^t \sqrt{\sum_{x \in \Omega} \frac{(\nu_0(x) - \pi(x))^2}{\pi(x)} + \frac{3 \max_{x \in \Omega} \left| \frac{\nu_0(x)}{\pi(x)} - 1 \right|}{1 - \lambda}} \sqrt{\frac{\pi(S)}{\pi(\bar{S})}}.$$

2.3.6.2 *Logarithmic Sobolev inequalities*

Logarithmic Sobolev inequalities arise in this context from the algebraic study of the transition matrices of Markov chains.

Associated with a reversible random walk is its (weighted) transition graph: i.e. a graph, (V, E), with vertex set the state space of the Markov chain, and edge weights given by

$$Q(\alpha, \beta) = \pi(\alpha)\mathbf{P}\left(\nu_{t+1} = \beta \mid \nu_t = \alpha\right) = \pi(\beta)\mathbf{P}\left(\nu_{t+1} = \alpha \mid \nu_t = \beta\right).$$

The logarithmic Sobolev constant of this weighted graph is then given by:

$$\alpha = \inf_{f \neq 0} \left\{ \frac{\sum_{(x,y) \in E} \left[(f(x) - f(y))^2 Q(x,y) \right]}{\sum_{x \in V} \left[f^2(x)\delta(x) \log \left(\frac{f^2(x) \sum_{z \in V} \delta(z)}{\sum_{z \in V} f^2(z)\delta(z)} \right) \right]} \right\},$$

where $\delta(z)$ is the degree of z. (This formula is drawn from Chung [29].)

Logarithmic Sobolev inequalities were initially applied to discrete finite Markov chains by Diaconis and Saloff-Coste [36]. They consider random walks on regular graphs, and show that in general the mixing time is bounded above by

$$\tau \leq \frac{1}{2\alpha} \log\log|\Omega| + \frac{2}{\lambda_2}.$$

Chung [29] extends a slightly stronger result to non-regular graphs.

The obvious trouble with using the logarithmic Sobolev inequality is that the formula for the log-Sobolev constant is exceptionally unwieldy. It has thus mainly been applied to cases where the Markov chain is either highly symmetric, or differs only slightly from a Markov chain that is highly symmetric.

The fact that the technique currently requires a high degree of symmetry should not necessarily be taken as an indication that the technique will prove fruitless in proving rapid mixing results: the coupling technique was held in a similar light in the not too distant past; Jerrum [66] wrote

> [The] "coupling" argument ...is very successful in the context of
> highly symmetric Markov chains ..., but seems very difficult or even im-
> possible to apply to the kind of Markov chains which arise in the analysis
> of Monte Carlo algorithms.

Diaconis and Saloff-Coste [36] have applied the log-Sobolev technique to show better mixing times for exclusion processes on the line and the finite circle than were previously known. They also used the technique to show rapid mixing of a random walk on a hyper-cubic portion of the n-dimensional lattice.

Frieze and Kannan [49] have used log-Sobolev inequalities to improve the bound on the mixing time for a Markov chain on a log-concave distribution on $\mathbb{R}^n$ (see also Section 4.3 in this book). This was an improvement on a bound on the mixing time of Frieze, Kannan, and Polson [50] that relied on a a Poincaré-type inequality, and of Applegate and Kannan [8] that relied on conductance techniques.

2.3.7 Coupling from the Past

This is a very useful idea that differs from all of the other Markov chain method approaches that we have looked at. Instead of using an analytical technique to find upper bounds on the mixing and relaxation times, utilization of this technique does not require *a priori* computation of these times, and is instead an algorithmic technique that calculates when to terminate itself. Additionally, the samples generated by this technique are exact rather than approximate.

The downside of this technique is that whilst powerful, it requires significant (even exponential) additional computational overhead. Fortunately this overhead can be mitigated almost entirely in some important cases.

The idea itself is elegantly simple, and is due to Propp and Wilson [100] (see also [101]). Despite the simplicity of the technique, the concept that underlies it propounds an unusual departure from the ways we ordinarily think of analyzing Markov chains.

Consider a Markov chain, $\mathcal{M}$, with stationary distribution π from which we want to sample exactly. Suppose we have one instance of this Markov chain $\mathcal{M}_\pi$ that was started at time $-\infty$, and as a consequence, has the stationary distribution for every finite time.

Suppose we can generate a *random map* from Ω to itself, chosen in such a way that, for any individual state, the random map describes one step of the Markov chain. Ideally we will want this random map to be constructed in such a way as to bring about coupling of any pair of states quickly.

Notionally, suppose we have such a random map for each time step from -1 to $-\infty$, which we shall call, $f_{-1}, f_{-2}, \ldots$. Define $F_{-t} = f_{-1} \circ f_{-2} \circ \cdots \circ f_{-t}$.

If there is a finite t, such that F_{-t} is a constant function, then under appropriate conditions that constant is an exact sample from the stationary distribution of $\mathcal{M}$. Somewhat less than formally, this can be seen by considering an instantiation of $\mathcal{M}_\pi$: as we have already asserted, this will have the stationary distribution for all time; yet, if it evolves under the random maps, then irrespective of its state at time $-t$, its state at time 0 is known, and is the constant that F_{-t} maps to.

The following two technical conditions must be adhered to for this scheme to work; they are drawn directly from Propp and Wilson [101].

1. The random map must be drawn from a distribution that preserves π, i.e. for a random map f

$$\sum_{i \in \Omega} \pi(i) \mathbf{P}(f(i) = j) = \pi(j) \quad \text{for all } j.$$

2. When $f_{-1}, f_{-2}, \ldots,$ are independent random maps, there is a non-zero probability that F_{-t} converges to a constant-valued function (i.e. there exists an N such that for all $n \geq N$, $F_{-n} = F_{-N}$).

This in essence, describes the coupling from the past technique. Clearly there are significant algorithmic problems with a facile implementation of such an algorithm, since one could envisage a scenario in which one would need to explicitly store the entire random map: and for an exponentially large state space, this alone would prevent the algorithm from being fast.

Fortunately there are many cases in which such problems may be avoided, e.g. in monotone Markov chains [100], anti-monotone Markov chains [61], and Markov chains on product spaces [64]. This is a very rapidly growing area of research, and an excellent guide is due to Wilson [129].

Chapter 3

Approximate Counting

Why do we want to count? Faced with this somewhat flippant question most people would splutter before saying that it would be in order to know how big something was. And flippant though the question is, and trivial though the answer, that really is the most salient reason.

Historically, and in wider life, there is hardly a single area where counting is not important. How large assets and income are form the essential basis for taxation; how large an army is has shaped the very countries in which we live. Much classical research has been concerned with geometry (see [40]), in particular areas and volumes—these problems too are a type of counting problem.

One common thread in all of these applications is that, usually, exact counting is not required and an approximate count is sufficient.

We will typically be dealing with counting problems that are #P-complete or #P-hard. Since we are working under the assumption that no tractable algorithm exists for an NP-complete problem, and these counting problems are harder even than NP-complete problems, it is clear that we cannot expect to find tractable *exact* solutions to our counting problems.

We need therefore to relax the answer that we demand. There are two relaxations that we consider. We can relax our demand that our answer be exact, and we can relax our insistence that our algorithms are deterministic, so as to allow randomized algorithms.

If we allow randomized algorithms, but still insist that we obtain an exact answer with some constant probability greater than one-half, we could, by taking the median of a sufficiently large number of executions of this algorithm, raise the probability of obtaining the exact answer to at least $(1 - \delta)$ for any $\delta > 0$. By using the defining reduction of the #P-complete problem, we could thus bootstrap a polynomial-time randomized algorithm to work for *any* #P-complete problem, including the counting variant of an NP-complete problem, e.g. #SAT. But this would mean that with a probability greater than one-half, we could distinguish formulæ that had no satisfying assignment from those that did, i.e that $NP \subseteq BPP$, which contradicts our assumption. Thus, performing just this relaxation is not enough.

We can instead relax the demand that the answer we obtain is exact. There are various relaxations that can be made, principally splitting into those that allow additive, and those that allow multiplicative errors. There has been no significant research on the effect of allowing additive errors that we are aware of; research following Stockmeyer [117] and Karp and Luby [78] has concentrated exclusively on multiplicative errors. Whilst it would no doubt be interesting to consider the effects of additive error we will not do so in this book, and we will henceforth concentrate solely on multiplicative error.[1]

If we allow a multiplicative error, yet still insist that the algorithm be deterministic, it *may* be possible to achieve our goal sometimes. There are currently no tractable deterministic approximation algorithms for #P-complete problems, nor are there any proofs that the existence of such an algorithm would force a collapse of any complexity classes. In some circumstances, though, it is provable that no such deterministic approximation algorithm exists, yet a randomized algorithm does (see Dyer and Frieze [40] for a discussion of this for the problem of approximating the volume of a convex body).

A *randomized approximation scheme* for Ω is a randomized algorithm which takes as input a (suitable encoding of) Ω and an error parameter ε, and outputs a number Y, such that $|\Omega|/(1+\varepsilon) \leq Y \leq |\Omega|(1+\varepsilon)$, with probability at least $3/4$. A randomized approximation scheme is said to be *fully polynomial* if it runs in time that is bounded (above) by a polynomial in the size of its input, and ε^{-1}. In fact it may be shown that instead of $3/4$, any inverse polynomial greater than $1/2$ will suffice [71]; we use $3/4$ merely by convention.

It is now common to refer to a fully-polynomial randomized approximation scheme as an *fpras* (pronounced "EF-PRAZ"), and we shall adopt this nomenclature from here on.

3.1   Parsimonious Reductions

We have commented already that we will be concerned primarily with problems that are #P-hard. It will be fruitful to pause for a moment to consider a central concept in #P-hardness proofs: parsimonious reductions.

Suppose we can transform a counting problem A to a counting problem B in polynomial time, in such a way that there is a bijection between solutions of A and B. Then, such a transformation is said to be a *parsimonious reduction* from A to B. Thus, if A is a #P-hard problem, then B is a #P-hard problem too. Similarly, if B has a polynomial-time algorithm, then A has one too.

It should be clear that such a definition is too restrictive for our purpose (see, for example, the proof of Lemma 29 on page 117); fortunately, Kozen [82] has defined a *counting reduction*. Suppose we can transform a counting problem A to a counting problem B in polynomial time, and that there is a polynomial-time computable function $t : \mathbb{N} \mapsto \mathbb{N}$, such that the number of solutions to A may be found by transforming

---

[1]Cai and Hemachandra [28], consider a different type of approximate counting, and show that unless the polynomial hierarchy collapses completely, i.e. P = PH, there is no polynomial-time algorithm for producing a short list of $n^{1-\epsilon}$ potential solutions, one of which is guaranteed to be correct, for #SAT.

to B, counting the number of solutions of B, and mapping this number through t. Then A is said to have a counting reduction to B, and the reduction is called *weakly parsimonious*. Then, as with parsimonious reductions, if A is a #P-hard problem, then B is a #P-hard problem too. Similarly, if B has a polynomial-time algorithm, then A has one too.

We will introduce a third type of reduction here, which we name an *almost parsimonious reduction*, which is motivated from randomized approximate counting, and is even weaker than a weakly parsimonious reduction.

A reduction from counting problem A to counting problem B is said to be an almost parsimonious reduction if

1. there is a randomized polynomial-time transformation from A to B,

2. there is an *fpras* for the mapping of the number of solutions of B to the number of solutions of A.

Thus, if A is almost-parsimoniously reducible to B, and there is an *fpras* for B, then there is an *fpras* for A.

Remark. Of course one could weaken the definition of an almost parsimonious reduction even further to allow the transformation from A to B to be from A to a random problem instance B' with some appropriate distribution, but investigating such reductions is outside the scope of this book. $\bigcirc$

3.2 Counting Directly

One of the simplest methods of approximately counting the number of elements of a set is if it is possible to sample from a distribution over the set and take the expectation of a relevant function over that distribution. Then a simple Monte Carlo procedure will yield an approximation. We consider a straightforward application of this idea in Section 3.2.1.

A significant limitation of the straightforward application of this type of technique, however, is the range of applications for which a direct sampling *fpras* is a possibility. We will see in Section 3.2.2 that this limitation is mitigated partly by the Karp–Luby technique.

3.2.1 Direct Sampling

Suppose we can embed our target set into some larger, and possible simpler set of structures. Suppose further, that the expectation of a suitable function on this larger set over some appropriate distribution is precisely the number we desire: the solution to the original counting problem (cf. Section 2.2.1).

Then we immediately have an algorithm for approximate counting. Of course this may or may not be an *fpras*, but this is the trade-off for the simplicity of the algorithm.

Good examples of this technique may be found in Alon, Frieze and Welsh [6], who gave, for example, an *fpras* for the partition function of the ferromagnetic Potts model in dense graphs.

3.2.2 Karp–Luby Technique

We now describe another, more subtle technique for counting directly, that is in some sense a generalization of the direct sampling technique of the previous section. This technique was developed by Karp and Luby [78], who were the first to give (and thus show that it is possible to give) an *fpras* for a problem that is #P-hard.

Suppose we have a base set Ω, with equivalence relation $\sim$, and we wish to approximately count the number of equivalence classes, $\Omega/\sim$. The Karp–Luby technique gives a mechanism for doing this.

The Karp–Luby method is inspiringly simple. It defines independent identically distributed variables X_i, such that $\mathbf{E}(X_i) = |\Omega/\sim|$, and simply takes a sample mean of a sufficiently large number of such X_i. There are two methods for defining the X_i. In *Method 1*, we choose an element $x_i \in \Omega$ uniformly at random, and then set $X_i = |\Omega|/|[x_i]|$, where $[x_i]$ denotes the equivalence class that contains x_i. In *Method 2*, we assume that there is a canonical representative of each equivalence class, and that we can determine if an element is the canonical representative of its equivalence class; again, we choose an element $x_i \in \Omega$ uniformly at random, and if x_i is a canonical representative, set X_i to $|\Omega|$, otherwise set X_i to 0.

Karp and Luby prove that if $p = |\Omega/\sim|/|\Omega|$, then it is sufficient to take

$$\frac{9}{2}\frac{(1-p)}{p}\frac{1}{\varepsilon^2}\ln\frac{8}{3}$$

random variables X_i, to ensure that their mean gives a randomized approximation scheme.[2]

One can use this technique to approximately sample from a variety of problems, for it essentially lets us simplify the problem by fixing some part of the solution.

Despite its being one of the founding techniques in this area, it has had rather limited application so far, with the most notable examples being papers of Karger [77] and Gore *et al.* [56].

- Karp and Luby's paper [78] was focused on showing that approximate counting was possible for several problems, although, as we will see in Section 3.3, this implies that approximate sampling is possible. Specifically, they give applications to estimating the number of triangulated plane maps with a given number of vertices, estimating the cardinality of a union of sets, approximating #DNF, and estimating the (2-terminal) probability of network reliability[3].

- Karp, Luby, and Madras [79] gave methods for speeding up the *fpras* for #DNF.

- Karger [77] gave an *fpras* for the all-terminal network reliability problem.

---

[2]Karp and Luby also present a third method that has expected running time faster than Methods 1 and 2.
[3]More strictly for estimating the probability of network failure.

- Gore *et al.* [56] gave a quasi-polynomial-time[4] algorithm for sampling words from a context free grammar.

3.3 Counting and Sampling

The connexion between the problems of counting and sampling was first formalized by Jerrum, Valiant, and Vazirani [71], and then later by Sinclair and Jerrum [112]. They considered combinatorial structures which have a property known as *self-reducibility*, which was first studied by Schnorr [108]. In essence, a combinatorial structure is self-reducible if it may be constructed inductively from a constant number of smaller combinatorial structures. For example, in the case of SAT, consider a Boolean formula, F, in conjunctive normal form. The solutions of this may be partitioned into those that assign True and False to a particular variable; each of these partitions represents a shorter Boolean formula than the original, thus SAT is self-reducible.

The central results of Jerrum, Valiant, and Vazirani, and Sinclair and Jerrum were theorems that for a self-reducible structure, an *fpras* exists if and only if a fully-polynomial!approximate sampler does. Furthermore, if there is a polynomial-time randomized approximation scheme for the counting problem, which provides estimates accurate to within a factor of $1 + O(|x|^\alpha)$, where $|x|$ is the input size and α is *any* real constant, then there is an *fpras* too.

3.4 The Markov Chain Monte Carlo Method

Although results of Jerrum, Valiant, and Vazirani gave us a method of constructing an *fpras* from a fully-polynomial almost uniform sampler for self-reducible structures, in practice this result is rarely used in its full majesty. Instead, the *Markov chain Monte Carlo method* is used. This is frequently abbreviated to MCMC in the literature. This has the added advantage of being applicable even for structures that are not known to be self-reducible, e.g. in using procedures for sampling points from a convex body to approximate the volume of the convex body.

The essence of the Markov chain Monte Carlo method is simple: rather than estimate the size of a structure directly, one estimates the ratio of sizes between the structure of interest, and one that is slightly simpler. Repeat this estimation for the ratio between this slightly simpler structure and one that is simpler still, and so forth, until one ends up with a structure that is sufficiently simple that one can calculate (or estimate) its size directly. For example, consider the task of estimating the number of sink-free orientations of a graph (we consider this problem in detail in Section 4.2). Here, we might, for instance, consider the ratio of the number of sink-free orientations of a graph, and the number of sink-free orientations of another graph formed from the first by contracting an edge. To do this, we could set up an indicator variable with

---

[4]Quasi-polynomial-time here means $n^{O(\log n)}$.

mean the reciprocal of the appropriate ratio in the following way: pick a sink-free orientation of the smaller graph (almost) uniformly at random, and a random orientation for the edge we contracted, to form an orientation of the larger graph. Return a 1 if this is a sink-free orientation or a zero otherwise. If we repeat this procedure, we obtain estimates for a sequence of ratios, until our final graph has only one vertex per component. We may calculate the number of sink-free orientations of this graph directly, and by multiplying this by all of the (approximated) ratios, we obtain an approximation to the number of sink-free orientations of the original graph.

We thus frequently find ourselves in the following scenario when reducing approximate counting to approximate sampling. Suppose $\mathcal{M}$ is a Markov chain on Ω, with stationary distribution the uniform distribution on Ω. If S is some subset of Ω, then we seek to estimate $\mu = |S|/|\Omega|$. Let I be the indicator function on Ω that equals 1 on elements of S, and zero otherwise. Thus $\mu = \sum_{\omega \in \Omega} I(\omega)/|\Omega|$.

There are two ways of proceeding. As usual, we will write $\tau_1(\varepsilon)$ for the mixing rate of $\mathcal{M}$, and τ_2 for its relaxation time. The naïve way would be to take s successive independent simulation runs of $\mathcal{M}$ each of length $\tau(\varepsilon_1)$, so as to generate almost uniform samples. We could then take the average of I over these approximate samples. Let us call these samples $Z_1, Z_2, \ldots, Z_s$, with sample average Z. Then (assuming $\varepsilon_2 > \varepsilon_1$)

$$\mathbf{P}(|Z - \mu| > \varepsilon_2) \leq \mathbf{P}(|Z - \mathbf{E}(Z)| + |\mathbf{E}(Z) - \mu| > \varepsilon_2).$$

But, from the definition of variation distance $|\mathbf{E}(Z) - \mu| \leq \varepsilon_1$, thus

$$\mathbf{P}(|Z - \mu| > \varepsilon_2) \leq \mathbf{P}(|Z - \mathbf{E}(Z)| > \varepsilon_2 - \varepsilon_1).$$

Thus, by Chebychev's inequality

$$\begin{aligned}
\mathbf{P}(|Z - \mu| > \varepsilon_2) &\leq \frac{\operatorname{Var} Z}{(\varepsilon_2 - \varepsilon_1)^2} = \frac{\sum_{i=1}^{s} \operatorname{Var} Z_i}{s^2(\varepsilon_2 - \varepsilon_1)^2} \\
&= \frac{\mathbf{E}(Z)(1 - \mathbf{E}(Z))}{s(\varepsilon_2 - \varepsilon_1)^2}.
\end{aligned}$$

Recalling that $|\mathbf{E}(Z) - \mu| \leq \varepsilon$ will help us to bound the right-hand side of this inequality, for particular values of μ. (A uniform bound can of course be obtained by noting that $\mathbf{E}(Z)(1 - \mathbf{E}(Z)) \leq 1/4$.)

In order to ensure that the right-hand side is bounded above by a constant, one can of course vary both s and ε_1. If we bound $\mathbf{E}(Z)(1 - \mathbf{E}(Z))$ by $1/4$ then we should be able to calculate the optimum value to take for these: the number of simulation steps we require will be approximately (for some constant k)

$$\frac{\tau}{k\varepsilon_2^2} \frac{\ln(1/\varepsilon_1)}{(1 - \varepsilon_1/\varepsilon_2)^2}.$$

As may easily be checked, this is minimized when $x = \varepsilon_1/\varepsilon_2$ satisfies $x^2 - x = 2(\ln x + \ln \varepsilon_2)$, so for small ε_2, we should have $\varepsilon_1 \approx 0.285\varepsilon_2$.

In conclusion, we may estimate μ, with additive error at most ε with probability at least $3/4$ in time approximately $0.652\tau \ln(1/\varepsilon)/\varepsilon^2$.

However, if we take $\varepsilon_1 = \varepsilon_2/2$, then the RHS is bounded above by $1/s\varepsilon_2^2$; i.e. we need to make $O(\tau(\varepsilon_2)/\varepsilon_2^2) = O(\tau \log(1/\varepsilon_2)/\varepsilon_2^2)$ simulation steps, in order to bound the RHS by a constant.

Recall that in the Markov chain Monte Carlo method, we have to estimate a sequence of ratios — say k ratios. Notionally, we seek to estimate ratios $\xi_1, \xi_2, \ldots, \xi_k$, using the Markov chain method for each. The naïve method for doing this is to generate independent approximate samples from an appropriate Markov chain. Suppose that the expected ratio under this approximate distribution is μ_i, and we obtain estimators $\bar{Z}_i$ for the μ_i by taking s_i independent indicator random variables, as discussed previously. How large do we need to take the s_i in this case though? We can certainly improve slightly on the estimates above.

MARKOV CHAIN MONTE CARLO LEMMA (LEMMA 5). *Suppose $\varepsilon \in (0,1]$. Suppose also that for all $i \in \{1, 2, \ldots, k\}$, we have $\xi_i \in (0, 1]$, and Z_i, which are indicator random variables with mean μ_i, such that $\xi_i/(1 + \varepsilon(2 - \varepsilon)/4k) \leq \mu_i \leq \xi_i(1 + \varepsilon(2 - \varepsilon)/4k)$. We may obtain estimators $\bar{Z}_i$, for μ_i, by taking the mean of s_i independent instances of Z_i, where the s_i are such that for all i we have $(1 - \mu_i)/\mu_i s_i \leq \gamma\varepsilon^2/k$, where $\gamma = \ln((11 - 2\sqrt{2})/8) \geq 1/48$. Then $(1+\varepsilon)^{-1}\xi_1\xi_2 \ldots \xi_k \leq \bar{Z}_1\bar{Z}_2 \ldots \bar{Z}_k \leq (1+\varepsilon)\xi_1\xi_2 \ldots \xi_k$, with probability at least $3/4$.*

Proof. By the definition of variance, observe that for all i, we have $(\mathrm{Var}\,\bar{Z}_i)/\mu_i^2 = (1 - \mu_i)/\mu_i s_i \leq \gamma\varepsilon^2/k$. We therefore have

$$\frac{\mathrm{Var}\,\bar{Z}_1\bar{Z}_2 \ldots \bar{Z}_k}{(\mu_1\mu_2 \ldots \mu_k)^2} = \prod_{i=1}^{k}\left(1 + \frac{\mathrm{Var}\,\bar{Z}_i}{\mu_i^2}\right) - 1 \leq \left(1 + \frac{\gamma\varepsilon^2}{k}\right)^k - 1 \leq e^{\gamma\varepsilon^2} - 1.$$

By Chebychev's inequality:

$$\mathbf{P}\left(\left|\bar{Z}_1\bar{Z}_2 \ldots \bar{Z}_k - \mu_1\mu_2 \ldots \mu_k\right| > \left(1 - \frac{1}{\sqrt{1+\varepsilon}}\right)\mu_1\mu_2 \ldots \mu_k\right) \leq$$

$$\frac{(1+\varepsilon)\left[e^{\gamma\varepsilon^2} - 1\right]}{2 + \varepsilon - 2\sqrt{1+\varepsilon}}.$$

It may be verified by elementary calculus that the right-hand side is monotone increasing with ε over the range $(0, 1]$, thus it is clear by substitution that the right-hand side is bounded above by $1/4$.

We therefore have that with probability at least $3/4$:

$$\frac{1}{\sqrt{1+\varepsilon}}\mu_1\mu_2 \cdots \mu_k \leq \bar{Z}_1\bar{Z}_2 \ldots \bar{Z}_k \leq \sqrt{1+\varepsilon}\mu_1\mu_2 \ldots \mu_k.$$

Note that $(1 + \varepsilon(2 - \varepsilon)/4k)^k \leq \sqrt{1+\varepsilon}$, since

$$(1 + X)^k \leq \sqrt{1+\varepsilon}$$

$$\Leftrightarrow \ln(1 + X) \leq \frac{1}{2k}\ln(1 + \varepsilon)$$

but $\ln(1 + X) \leq X$, and $\varepsilon - \varepsilon^2/2 \leq \ln(1 + \varepsilon)$, so this is implied by

$$X \leq \varepsilon(2 - \varepsilon)4k.$$

Thus, since $\xi_i/(1 + \varepsilon(2 - \varepsilon)/4k) \leq \mu_i \leq \xi_i(1 + \varepsilon(2 - \varepsilon)/4k)$, we have

$$\frac{1}{1 + \varepsilon}\xi_1\xi_2\ldots\xi_k \leq \bar{Z}_1\bar{Z}_2\ldots\bar{Z}_k \leq (1 + \varepsilon)\xi_1\xi_2\ldots\xi_k,$$

with probability at least $3/4$. $\square$

Note that the constants in the above lemma were chosen for convenience of proof rather than for optimality.

Remark. In applying the Markov chain Monte Carlo Lemma, we will need to calculate the s_i. The bound in the lemma is given as $(1 - \mu_i)/\mu_i s_i \leq \gamma\varepsilon^2/k$, or rearranging

$$s_i \geq \left(\frac{1}{\mu_i} - 1\right)\frac{k}{\gamma\varepsilon^2}.$$

Thus to gain a lower bound on s_i, we will need a lower bound on μ_i. We find in practice that the only bound we have on μ_i, however, is given by our bound on ξ_i. Substituting this in and rearranging, we see that

$$s_i \geq \frac{k}{\gamma\varepsilon^2}\left(1 + \frac{1}{\xi_i}\right) + \frac{2 - \varepsilon}{4\varepsilon\xi_i\gamma},$$

or throwing away some tightness in exchange for a simpler inequality:

$$s_i \geq \frac{k}{\gamma\varepsilon^2}\left(1 + \frac{1 + 1/4k}{\xi_i}\right).$$

                                                                                            $\bigcirc$

Another method for estimating μ is due to Aldous [2], although a clearer exposition is given in Aldous and Fill [3]. Here, one considers a single simulation run of $\mathcal{M}$, of expected length $O(\tau + \varepsilon_2^{-2}\tau_2)$, and considers the average of I over all states of the Markov chain from step $\Theta(\tau)$ to the end of the simulation run; it transpires that this suffices to ensure that $\mathbf{P}(|Z - \mu| > \varepsilon_2)$ is bounded above by $1/4$.

Chapter 4

Applications: Coupling

In this chapter we consider three non-trivial applications of the coupling method to show rapid-mixing of certain Markov chains. In each case, the associated exact counting problem is #P-complete, but the approximate counting problem succumbs to the Markov chain Monte Carlo Lemma, once rapid-mixing of an appropriate Markov chain has been established.

In Section 4.1, we extend a result of Jerrum [67] on graphs to hypergraphs. Jerrum gave an *fpras* for the number of k-colourings of a graph, provided k is more than twice the maximum degree of the graph. Not only do we extend this result to hypergraphs, we show how the result extends also to the case where k is equal to twice the maximum degree of the graph.

In Section 4.2, we consider the combinatorial problems associated with graph orientations with no sink, and the associated problem of TWICE-SAT, which is the familiar SAT problem restricted to the case where each variable appears at most twice.

In Section 4.3, we consider a problem that has been considered by many authors previously: that of approximately uniformly sampling a point from within a convex body, and the counting variant, which is to approximate the volume of the convex body. Whilst Markov chains have been shown to be rapidly mixing for this class of problems by other authors, this is the first analysis of such a class of Markov chains that uses the coupling method: previous methods have relied on conductance, log-Sobolev inequalities, or Poincaré inequalitites. This section also differs from the other sections in this chapter, and from the applications in Chapter 5, in that it is the only application we consider for which the Markov chains have continuous state spaces. There are some additional technical considerations that this engenders, but these are dealt with *in situ*, so as not to obfuscate the general techniques used elsewhere in this book.

4.1   Hypergraph Colourings

4.1.1   Introduction

Suppose G is a hypergraph with vertex set V and edge set $E \subseteq \mathbb{P}(V)$. A k-colouring
of G is a function $\chi : V \to C$ (where C is a set of k colours) satisfying the property
that no edge with more than one vertex is monochromatic[1] (i.e. contains vertices of
only one colour). This is a generalization of the familiar concept of a k-colouring of a
graph. Exact enumeration of the number of k-colourings of a graph (and thus hypergraphs) is a well-known #P-complete problem (see e.g. [44]), so it is unlikely that any
randomized polynomial-time algorithm for its exact computation exists. Indeed, for
fixed $k > 2$, there cannot be an *fpras* for the number of k-colourings of a graph unless
$\mathrm{NP} = \mathrm{RP}$ (see below).

If we impose additional constraints on the graph, however, the situation changes:
Jerrum [67] has exhibited an *fpras* for the number of k-colourings of a graph, provided
that k is larger than twice the maximum degree of the graph. In this section, we
generalize Jerrum's result to hypergraphs, and *inter alia*, extend it to the case where k
is precisely twice the degree of the graph. It is easy to see that even in this restricted
case, the problem remains #P-complete, for, given a polynomial-time algorithm for
calculating exactly the number of k-colourings of a graph for distinct values of k, we
could use Lagrange interpolation to calculate the chromatic polynomial of a graph,
which is a #P-hard problem (see [44]).

The *fpras* described here uses the Markov chain Monte Carlo method, in the same
fashion as Jerrum's *fpras*. In essence, given an initial k-colouring, a (rapidly mixing)
Markov chain on the set of all k-colourings is simulated for sufficiently long to provide
a k-colouring generated almost uniformly at random. This is then used to estimate the
ratio of the number of k-colourings between G, and G with one of its edges deleted.
Repeating this until all edges are deleted, we gain our desired approximation by taking
the product of these ratios together with the (immediately calculable) number of k-
colourings of n isolated vertices.

Hypergraph colouring problems are, in some sense, harder in complexity terms
than their graph colouring counterparts: e.g. deciding if a hypergraph is 2-colourable
is an NP-complete problem, and counting the number of 2-colourings is a #P-complete
problem, whereas both deciding if a graph is 2-colourable and counting the number of
2-colourings are trivially polynomial.

Work has nonetheless progressed on some specific cases of counting hypergraph
colourings, e.g. Berge has considered complete r-partite hypergraphs [14].

In Section 4.1.2, we define a Markov chain with state space the set of colourings
of a hypergraph. We go on to show that, provided k is sufficiently large, this Markov
chain is rapidly mixing and has uniform stationary distribution. The precise conditions
for k to be "sufficiently large" are explained in Sections 4.1.2.3 and 4.1.2.4. Finally,

---

[1] This is sometimes known as a *weak* k-colouring; a *strong* k-colouring satisfies the property that no two
vertices in the same edge have the same colour, and as a computational problem is thus reducible to graph
colouring.

in Section 4.1.3 we show how the simulation of this Markov chain may be used as a basis for an *fpras* for hypergraph colourings.

4.1.1.1 Notation and preliminaries

As edges with only one vertex have no proper colouring, we shall assume for the remainder of Section 4.1 that the hypergraphs under consideration do not have any of these singleton edges. Furthermore, we may also assume that no edge is a superset of another in the hypergraphs we consider, for this does not affect the number of colourings. We will also make the assumption that the hypergraphs under consideration are *connected*.

Define $\Omega_k(G)$ to be the set of all k-colourings of G. The number of vertices and edges of G will be denoted $n = n(G)$, and $m = m(G)$ respectively.

For the hypergraphs under consideration, the definition of a k-colouring is standard, and may be found, for example, in Berge [15] or Tomescu [120].

We will say that a colour, c, is *critical* for a vertex, v, in a given k-colouring if re-colouring v with c would not result in a k-colouring. We will say that an edge is critical for v if all vertices in it except for v are coloured with the same (critical) colour, and that colour is the critical colour of the edge for that vertex.

Observe that the intersection of any pair of edges critical with different colours for a vertex, v, must be precisely $\{v\}$, and thus from the definition of δ_1 (see Section 1.4.2), an upper bound on the number of critical colours that v may have is $\delta_1(v)$.

4.1.2 Approximate Sampling

In this section, we provide a *fully-polynomial almost uniform sampler* for k-colourings of a hypergraph, for k sufficiently large.

Consider the homogeneous Markov chain $\mathcal{M}(G, k)$ with time states (X_t), state space $\Omega_k(G)$ and transition probabilities modelled by:

1. choose a vertex, $v \in V$ and a colour, $c \in C$ uniformly at random.

2. suppose $X_t = \chi$; define χ' by $\chi'(w) = \chi(w)$ for all $w \neq v$, and $\chi'(v) = c$. If χ' is a k-colouring of G then let $X_{t+1} = \chi'$, otherwise let $X_{t+1} = X_t = \chi$.

LEMMA 6. *If $k \geq \Delta_1 + 2$ then $\mathcal{M}(G, k)$ has a unique stationary distribution, and furthermore this is the uniform distribution.*

Proof. Suppose α and β are arbitrary states of $\mathcal{M}(G, k)$ (i.e. k-colourings of G). Note that α communicates with β, since there is a strictly positive probability of α being transformed into β. One such transformation is thus: suppose c is a colour in β. Then for each vertex, v, in α of colour c, we could re-colour v to a different colour, since there are at most Δ_1 critical colours for v, and we have assumed $k \geq \Delta_1 + 2$. Then each vertex in α that corresponded to a c-coloured vertex in β could then be re-coloured to c. This could then be repeated with every other colour in β. Therefore, since α and β were arbitrary states, $\mathcal{M}(G, k)$ is irreducible.

From the definition of $\mathcal{M}(G, k)$, it is easily seen that for all states, there is a positive probability of the state remaining unchanged after a further time-step, since the identity transition always has positive probability. It follows therefore that $\mathcal{M}(G, k)$ is aperiodic.

Since $\mathcal{M}(G, k)$ is irreducible and aperiodic, it is ergodic, and, from the detailed balance equation, (1.2), it follows that its stationary distribution is uniform. $\square$

An almost uniform sampler, therefore, could be an algorithm that found any k-colouring of G, and, using that as an initial state, simulated $\mathcal{M}(G, k)$ for a sufficiently long time to bring the probability distribution of the state of the Markov chain to within a variation distance of ε from its uniform stationary distribution. The sampler will be fully polynomial if $\mathcal{M}(G, k)$ is rapidly mixing.

Finding an initial state is easy to do in polynomial time, provided that $k \geq \Delta_1 + 1$. Initially, assume that all vertices are uncoloured. Let us assume that the colours have been enumerated $c_1, c_2, \ldots, c_k$, and the vertices $v_1, v_2, \ldots, v_k$. Take the first colour, and successively cycle through each of the uncoloured vertices, colouring them if doing so does not create a monochromatic edge. Repeat with the second and subsequent colours. That this algorithm will terminate with all vertices having been coloured is fairly easy to see, for suppose otherwise, and after this procedure has been carried out with all $\Delta_1 + 1$ colours, there is a vertex v that remains uncoloured. Then v must be critical for every colour, else it would have been coloured by the procedure. But as we have noted already, $\delta_1(v)$ is an upper bound on the number of critical colours a vertex may have; this contradiction establishes the correctness of the algorithm. This algorithm is clearly polynomial-time, and is due to Tomescu [120].

We now show that the Markov chain $\mathcal{M}(G, k)$ is rapidly mixing for sufficiently large k, using the coupling method.

We therefore describe a coupling of X_t and Y_t, each of which in isolation is a copy of $\mathcal{M}(G, k)$, such that there is a (probabilistic) tendency for the two Markov chains to couple.

The coupling we use is equivalent to one in [67] and has transitions modelled by the following:

1. choose a vertex $v \in V$ and a colour, $c \in C$ uniformly at random.

2. compute a permutation g of C, using the procedure described below in Section 4.1.2.1 (this choice of permutation is independent of the choice of c).

3. re-colour vertex v with colour c in X_t if this results in a k-colouring, otherwise make the identity transition, and similarly, re-colour vertex v with colour $g(c)$ in Y_t if this results in a k-colouring, making the identity transition otherwise.

Note that so long as the computation of g is independent of the choice of c, the distribution of $g(c)$ is uniform, hence both (X_t) and (Y_t) are true instances of $\mathcal{M}(G, k)$.

4.1.2.1 Computing the permutation

Let $A_t \subseteq V$ be the set of vertices on which X_t and Y_t agree, and $D_t \subseteq V$ the ones on which they disagree.

Define $C_X(v) \subseteq C$ to be the set of critical colours for v in X_t. Define $C_Y(v)$ analogously.

Now g is computed as follows:

- if $v \in D_t$ then g is the identity.

- if $v \in A_t$ then suppose, without loss of generality that $|C_X(v) \setminus C_Y(v)| \leq |C_Y(v) \setminus C_X(v)|$. Choose any subset $C'_Y(v) \subseteq C_Y(v) \setminus C_X(v)$ with $|C'_Y(v)| = |C_X(v) \setminus C_Y(v)|$. Let m be an arbitrary bijection from $C'_Y(v)$ to $C_X(v) \setminus C_Y(v)$. Then

$$g(c) = \begin{cases} m(c) & \text{if } c \text{ is in the range of } m, \\ m^{-1}(c) & \text{if } c \text{ is in the domain of } m, \text{ and} \\ c & \text{otherwise.} \end{cases}$$

4.1.2.2 Rapidity of coupling

We consider next how $|D_t|$ evolves over time. Clearly, when $|D_t| = 0$ the random walks will have coupled.

Consider L_t, the probability that $|D_{t+1}| = |D_t| + 1$. This occurs if and only if the vertex chosen $v \in A_t$, and the colour subsequently chosen is a member of $C_Y(v) \setminus C_X(v)$, or $C_X(v) \setminus C_Y(v)$ that maps to itself under g (if any other colour is chosen, then the two chains will either both re-colour vertex v, or both will stay the same), i.e.

$$L_t = \mathbf{P}\left(|D_{t+1}| = |D_t| + 1\right) = \frac{1}{kn} \sum_{v \in A_t} \left||C_Y(v) \setminus C_X(v)| - |C_X(v) \setminus C_Y(v)|\right|.$$

$$(4.1)$$

Consider now S_t, the probability that $|D_{t+1}| = |D_t| - 1$. This occurs if and only if the vertex chosen $v \in D_t$ and the colour chosen is not the colour of a critical edge in either X_t or Y_t — the number of colours which satisfy this is $k - |C_X(v) \cup C_Y(v)|$. Hence

$$S_t = \mathbf{P}\left(|D_{t+1}| = |D_t| - 1\right) = \frac{1}{kn} \sum_{v \in D_t} \left(k - |C_X(v) \cup C_Y(v)|\right). \qquad (4.2)$$

LEMMA 7. *If there is a constant, $a > 0$, that satisfies $a|D_t| \leq S_t - L_t$ for all t, pointwise, then the mixing time, τ, satisfies $\tau \leq \lceil a^{-1} \ln(en) \rceil$, and the mixing rate, $\tau(\varepsilon) \leq \lceil a^{-1} \ln(n\varepsilon^{-1}) \rceil$.*

Proof. We have that

$$\begin{aligned} \mathbf{E}\left(|D_{t+1}|\right) &= L_t(|D_t| + 1) + S_t(|D_t| - 1) + (1 - L_t - S_t)|D_t| \\ &= |D_t| + L_t - S_t. \end{aligned}$$

Thus $\mathbf{E}\left(|D_{t+1}|\right) \leq (1 - a)|D_t|$, and hence $\mathbf{E}\left(|D_t|\right) \leq (1 - a)^t |D_0| \leq n(1 - a)^t \leq ne^{-at}$. Furthermore, since $|D_t|$ is a non-negative integer random variable $\mathbf{P}(|D_t| > 0) \leq \mathbf{E}\left(|D_t|\right) \leq ne^{-at}$. However $\mathbf{P}(X_t \neq Y_t) = \mathbf{P}(|D_t| > 0)$. Applying Lemma 6 and the Coupling Lemma completes the proof. $\qquad \square$

4.1.2.3 Two sufficient conditions

This section is devoted to finding two sets of sufficient conditions on k to ensure the existence of a constant a, such that $0 < a|D_t| \leq S_t - L_t$, in order to satisfy Lemma 7.

A condition of $k > \Delta_1 + \Delta_3$ Let

$$\begin{aligned}
\eta &= |\{(u,v) \mid u \in A_t, v \in D_t, \text{ and } u \text{ and } v \text{ are adjacent}\}|, \text{ and}\\
\nu &= |\{(u,v) \mid u \in D_t, v \in D_t, \text{ and } u \text{ and } v \text{ are adjacent}\}|.
\end{aligned}$$

Now $L_t \leq \sum_{v \in A_t} |C_X(v) \oplus C_Y(v)| / kn$, where $\oplus$ denotes symmetric set difference. Observe that each edge that is critical for v in exactly one of X_t and Y_t must contain a vertex in $D_t \setminus \{v\}$. Since the number of colours that are critical for a vertex in exactly one of X and Y is no greater than the number of edges that are critical for a vertex in exactly one of X and Y, we may associate each such critical edge with a vertex in D_t contained within it. Observe that the critical edges of a vertex v must each be associated with distinct vertices of D_t. In other words, for each v, the set of edges critical in precisely one colouring for v has a transversal contained in D_t. It follows therefore that $L_t \leq \eta/kn$.

Recall $S_t = \sum_{v \in D_t} (k - |C_X(v) \cup C_Y(v)|) / kn$. We may rewrite this as $S_t = |D_t|/n - \sum_{v \in D_t} |C_X(v)|/kn - \sum_{v \in D_t} |C_Y(v) \setminus C_X(v)|/kn$. In a similar fashion to the above, for each vertex v in D_t, the set of edges critical in Y, but not in X, has a transversal contained in $D_t \setminus \{v\}$. Thus $\sum_{v \in D_t} |C_Y(v) \setminus C_X(v)| \leq \nu$. But $\eta + \nu \leq \sum_{v \in D_t} \delta_3(v) \leq |D_t|\Delta_3$. Recall also that $\sum_{v \in D_t} |C_X(v)| \leq \sum_{v \in D_t} \delta_1(v) \leq |D_t|\Delta_1$.

Collecting these inequalities, we thus obtain

$$S_t - L_t \geq |D_t| \left(\frac{k - \Delta_1 - \Delta_3}{kn} \right),$$

whence we obtain if $a = (k - \Delta_1 - \Delta_3)/kn$, then a sufficient condition for $0 < a|D_t| \leq S_t - L_t$ is to have $k > \Delta_1 + \Delta_3$.

In fact, this bound may also be obtained using a different method: for details of how to obtain this bound using Dobrushin's uniqueness criterion, see Appendix A.

A condition of $k > 2\Delta_2$ The approach here is to ensure that an edge containing more than two vertices is counted as being critical at most once for each of the two random walks.

Define R_t to be the maximal set of edges with the following properties: each edge intersects both A_t and D_t; each edge which contains more than one vertex of A_t colours all vertices that it shares with A_t, except one, in one colour, the other vertex being coloured in a different colour. Then define:

$$\begin{aligned}
m_1 &= |\{e \in R_t \mid |e \cap A_t| = 2\}|\\
m_2 &= |\{e \in R_t \mid |e \cap A_t| \neq 2\}|\\
r_1(v) &= |\{e \in R_t \mid |e \cap A_t| = 1; v \in e\}|\\
r_2(v) &= |\{e \in R_t \mid |e \cap A_t| > 1; v \in e\}|.
\end{aligned}$$

Recall equation (4.1):

$$L_t = \frac{1}{kn} \sum_{v \in A_t} \big||C_Y(v) \setminus C_X(v)| - |C_X(v) \setminus C_Y(v)|\big|.$$

On the right-hand side of this equation, we are counting the number of vertices in A_t for which we could choose a colour that would result in an increase in the size of D_{t+1}, and we are counting with multiplicity equal to the number of such colours for each vertex.

Suppose we choose a vertex v and a colour c that would result in increasing the size of D_t. Then we must have $v \in A_t$, and there must be an edge e that is critical for v in c in only one of X_t and Y_t; therefore e contains a vertex in D_t. It follows that $e \in R_t$. Let f be a mapping for each pair (v, c) that would result in increasing the size of D_t maps (v, c) to such an edge $e \in R_t$.

Consider now the number of vertex/colour choices that map to an edge $e \in R_t$ under f. If e does not contain two vertices from A_t, then e must be critical for only one vertex. If e contains more than two vertices from A_t, it can only be critical in one colour, and thus there is at most one vertex/colour choice that maps to e under f. If e contains only one vertex, v, from A_t, it is either critical in only one colouring for v, or in both. If it is critical in only one, then clearly there is at most one vertex/colour choice that maps to e under f; if it is critical in both then either the two critical colours are in $C_Y(v) \setminus C_X(v)$ and $C_X(v) \setminus C_Y(v)$ respectively, in which case there is no vertex/colour choice that maps to e under f, or one of the critical colours is in $C_Y(v) \setminus C_X(v)$ or $C_X(v) \setminus C_Y(v)$, and the other in $C_X(v) \cap C_Y(v)$, in which case there is at most one vertex/colour choice that maps to e under f. Finally, if e contains precisely two vertices from A_t, say v and w which are coloured respectively c and d (in both X_t and Y_t), then the only possible vertex/colour choices that could map to e are (v, d) and (w, c). Putting this together, we see that

$$L_t \leq \frac{2m_1 + m_2}{kn}. \tag{4.3}$$

Consider now, for $v \in D_t$, the expression $|C_X(v) \cup C_Y(v)|$. We shall bound this above by using inclusion/exclusion: for each of the critical colours there must be a critical edge, so we count all of these, once for each colouring, giving $2\delta_2(v)$. Now of these, those that are in R_t and have only one vertex in A_t are critical with the same colour in both colourings, so need only be counted once; also those that are in R_t with more than one vertex in A_t cannot, by the definition of R_t, be critical for any vertex in D_t, since the vertices in A_t cannot all share the same colour. Hence

$$|C_X(v) \cup C_Y(v)| \leq 2\delta_2(v) - r_1(v) - 2r_2(v). \tag{4.4}$$

So, by substituting inequality (4.4) into equation (4.2), we have

$$S_t \geq |D_t| \left(\frac{k - 2\Delta_2}{kn} \right) + \sum_{v \in D_t} r_1(v) + 2r_2(v).$$

If, for an edge $e \in R_t$, there is only one element in the intersection with A_t, then e is counted once by $2m_1 + m_2$, and $|D_t \cap e| \geq 1$ times by $\sum_{v \in D_t} (r_1(v) + 2r_2(v))$. Similarly, if an edge $e \in R_t$ has precisely two elements in common with A_t, then it is counted twice by $2m_1 + m_2$, and $2|D_t \cap e| \geq 2$ times by $\sum_{v \in D_t} (r_1(v) + 2r_2(v))$. Finally, if there are more than two elements in the intersection of an edge $e \in R_t$

with A_t, then e is counted just once by $2m_1 + m_2$, but $2|D_t \cap e| \geq 2$ times by $\sum_{v \in D_t} (r_1(v) + 2r_2(v))$. Thus $\sum_{v \in D_t} (r_1(v) + 2r_2(v)) \geq 2m_1 + m_2$. Putting this together with inequality (4.3) then yields

$$S_t - L_t \geq |D_t| \left(\frac{k - 2\Delta_2}{kn} \right), \tag{4.5}$$

whence if $a = (k - 2\Delta_2)/kn$ we see that a sufficient condition for $0 < a|D_t| \leq S_t - L_t$ is to have $k > 2\Delta_2$.

Although we do not make use of this fact here, it is worth noting that in the restriction of the problem to those hypergraphs where all edges contain at least 4 vertices, we may improve the inequality above to $\sum_{v \in D_t} (r_1(v) + 2r_2(v)) \geq 4m_1 + 2m_2$, which would result in a speed-up for the cases $k > 2\Delta_2$.

4.1.2.4 Almost uniform sampling for $k = 2\Delta_2$

Consider now the case where $k = 2\Delta_2$. We start off with a short lemma:

LEMMA 8. $S_t = \mathbf{P}(|D_{t+1}| = |D_t| - 1) > 0$ *unless we have coupled or G is a simple regular graph, and* $|D_t| = n$.

Proof. Recall that $|D_{t+1}| = |D_t| - 1$ if and only if the vertex chosen is in D_t, and the colour chosen is not in $C_X(v) \cup C_Y(v)$. Now, the maximum number of critical colours that a vertex may have in a colouring is bounded above by Δ_1. Since $\Delta_2 \geq \Delta_1$, we have immediately that each vertex v in D_t is covered by exactly Δ_1 edges, all of which are critical in both colourings for v. However, if an edge is critical for a vertex in D_t in both colourings, it must have a second vertex in D_t.

We have already noted in Section 4.1.2.3 that if an edge is critical for two of its vertices in a single colouring, then that edge must cover precisely two vertices. Hence all edges incident on a vertex in D_t must cover precisely two vertices, both of which are in D_t. We have by assumption that our graph is connected, so we have that G must be a simple regular graph and $|D_t| = n$. $\qquad\square$

Recall that in order to discover the length of time for which it is necessary to simulate the Markov chain in order to achieve almost uniform sampling, we must bound above $\mathbf{P}(|D_t| \neq 0)$. To do this we shall consider a homeomorphism from the coupling procedure, $W_t = W(X_t, Y_t)$. This will be defined as follows: if $\mathbf{P}(|D_{t+1}| = |D_t| - 1) \neq 0$ then $W_t = |D_t|$, otherwise $W_t = n + 1$. Note that W_t is not in general Markovian.

Although we cannot in any meaningful sense talk of the exact probabilities of how W_t evolves, we can establish some uniform bounds. It is transparent from the definition of the coupling, of W_t and Lemma 8 that the probability of W_t changing by more than one is zero; and from equation (4.5), and Lemma 8, that, for $0 < W_t < n$

$$\begin{aligned}
\mathbf{P}(W_{t+1} = W_t - 1) &\geq \mathbf{P}(W_{t+1} = W_t + 1), \text{ and} \\
\mathbf{P}(W_{t+1} = W_t - 1) &\geq 1/kn.
\end{aligned}$$

Indeed, as the next lemma shows, these inequalities hold even when $W_t = n$.

LEMMA 9.

1. If $W_t = n + 1$ then $\mathbf{P}(W_{t+1} = W_t - 1) = (k-2)/k$,

2. If $W_t = n$ then

(a) $\mathbf{P}(W_{t+1} = W_t - 1) \geq 1/kn$, and

(b) $\mathbf{P}(W_{t+1} = W_t - 1) \geq \mathbf{P}(W_{t+1} = W_t + 1)$.

Proof. Suppose $W_t = n + 1$, then we know from Lemma 8 that the graph is simple, regular, and each vertex assumes different colours under X_t, and Y_t. Without loss of generality, let us suppose that we have chosen a vertex v that we will try to re-colour to colour c. If c is one of the colours that v already assumes under either X_t or Y_t, then from the definition of W the coupled process makes the identity transition. If however, c is one of the $k - 2$ other colours, then one, of X_t or Y_t must be re-coloured, say from d (to c). If this occurs, then any neighbour of v could make the transition from being in D_{t+1} to being in A_{t+2}, if it were the next vertex chosen, and the choice of colour were d. This establishes the first part of the lemma.

For the second part of this lemma, the first inequality is immediate from Lemma 8. To avoid trivialities, we will assume that $\mathbf{P}(W_{t+1} = n + 1) > 0$. Observe now, from the definition of the coupling, that a step that would lead to $W_{t+1} = n + 1$ must be reversible, i.e. if $W_{t+1} = n + 1$ then there must exist a further step such that $X_{t+2} = X_t$ and $Y_{t+2} = Y_t$. Since the coupling process is Markovian, we may assume — and it will be convenient to do so — that $W_{t-1} = n+1$, and that the transition from (X_{t-1}, Y_{t-1}) to (X_t, Y_t) was accomplished by re-colouring vertex v from colour c to d in X.

Thus, in (X_t, Y_t), the neighbours of v must each be adjacent to two vertices that are coloured d in either X_t or Y_t, and no vertices that are coloured c. Since, in order for W_{t+1} to equal $n + 1$ each vertex must be adjacent to precisely one vertex for each colour, there are clearly at most 2 vertices and one colour that could be chosen that would bring this about, i.e. $\mathbf{P}(W_{t+1} = n + 1 \mid W_t = n) \leq 2/kn$. However if we instead chose any neighbour of v and re-coloured it d, then we would have $W_{t+1} = n - 1$. Since there are Δ such neighbours, and we have assumed $\Delta_2 > 1$, the result is established. $\qquad\square$

Suppose now that we were to modify the coupling slightly, by adding an additional probability of making the identity transition, scaling down the probabilities of existing transitions *pro rata*. Call this process (X', Y'). Clearly, if we were to do this, we would only be increasing the time for the processes to couple: without going into laborious details, if T_W, and T_V are the random variables denoting the coupling time of the old and new processes respectively, and f is any monotone increasing function, then $\mathbf{E}\left(f(T_V)\right) \geq \mathbf{E}\left(f(T_W)\right)$. It is clear that we could construct the process (X', Y') in such a way that we would ensure that $\mathbf{P}(W(X'_{t+1}, Y'_{t+1}) = W(X'_t, Y'_t) - 1) = 1/kn$, for $0 < W(X'_t, Y'_t) < n + 1$, and $\mathbf{P}(W(X'_{t+1}, Y'_{t+1}) = W(X'_t, Y'_t) - 1) = (k-2)/k$ for $W(X'_t, Y'_t) = n + 1$.

Consider now, a one-dimensional random walk, U, on $\{0, 1, \ldots, n+1\}$, that has an absorbing state at 0, and transition probabilities given by: if $0 < U_t < n + 1$, then the probabilities of increasing, or decreasing are both equal to $1/kn$, and if $U_t = n + 1$, the probability of U decreasing is $(k - 2)/k$. Comparing this random walk with (X', Y'), we see that the probabilities of decreasing are identical, and the probabilities of increasing are greater in U, whence $\mathbf{E}\left(f(T_U)\right) \geq \mathbf{E}\left(f(T_V)\right)$, where T_U is the random variables denoting the time for U to reach zero.

We see, therefore, if we analyse the probability that $T_U > \tau$ that this is an upper bound on the probability that our coupling process has not coupled by time τ.

It is straightforward to analyse the expected time for the random walk U to reach the zero state, using standard techniques on difference equations, and thence obtain

$$\mathbf{E}\left(T_U \mid T_0 = a\right) = \frac{k(2an^2 + an - a^2 n)}{2} + \frac{ka}{k - 2}.$$

In particular,

$$\mathbf{E}\left(T_U \mid T_0 = n + 1\right) = \frac{kn^3 + kn^2}{2} + \frac{kn + k}{k - 2}.$$

Then, using Markov's inequality, and using the fact that T_U dominates T_W, we have that

$$\mathbf{P}(T_W \geq t + 1) \leq \mathbf{P}(T_U \geq t + 1) \leq \frac{k}{t + 1}\left(\frac{n^3 + n^2}{2} + \frac{n + 1}{k - 2}\right).$$

As mentioned above, this provides us immediately with a result on our original random walk, $\mathcal{M}(G, k)$.

LEMMA 10. *The Markov chain, $\mathcal{M}(G, k)$, has mixing rate satisfying*

$$\tau(\varepsilon) = \left\lceil \ln(\varepsilon^{-1})\tau \right\rceil, \textit{ where } \tau \leq \left\lceil ek\left(\frac{n^3 + n^2}{2} + \frac{n + 1}{k - 2}\right) - 1 \right\rceil.$$

Proof. Using the application of Markov's inequality above, we have that, in order to ensure that the probability we have not coupled is no greater than e^{-1}, it suffices to simulate $\left\lceil ek\left(\frac{n^3 + n^2}{2} + \frac{n + 1}{k - 2}\right) - 1 \right\rceil$ steps of the random walk. Since we may run successive, independent coupling "trials" of length τ, to ensure that the probability that we have not coupled is no greater than ε, we need simulate no more than $\left\lceil \ln\left(\varepsilon^{-1}\right) \right\rceil \tau$ steps. Applying the Coupling Lemma completes the proof. $\qquad\square$

4.1.3 The Approximation Scheme

We use the Markov chain!Monte Carlo method. The essential points of the scheme are as follows: we construct a sequence of graphs with G as its first element, with each successive element being generated by the removal of an edge from the previous element. We then estimate the ratio of the number of k-colourings between successive graphs in the sequence.

Let $G_m = G$, and for each a, such that $0 \leq a < m$, define G_a to be the graph obtained by deleting an edge of G_{a+1}. This edge should be chosen so that there is at most one component of G_a that is not a singleton. Note that deleting edges cannot increase either Δ_1 or Δ_2. Then

$$|\Omega_k(G)| = \frac{|\Omega_k(G_m)|}{|\Omega_k(G_{m-1})|} \times \frac{|\Omega_k(G_{m-1})|}{|\Omega_k(G_{m-2})|} \times \cdots \times \frac{|\Omega_k(G_1)|}{|\Omega_k(G_0)|} \times |\Omega_k(G_0)|.$$

Clearly, $|\Omega_k(G_0)| = k^n$.

To avoid trivialities, we shall assume that $m \geq 1$, and $\varepsilon \in (0, 1]$. Let ℓ be a choice function $\ell : E \to V$. Suppose further that the graphs G_i and G_{i-1} differ in the edge $e = \{v_1, v_2, \ldots, v_r\}$. Note that each colouring in $\Omega_k(G_{i-1}) \setminus \Omega_k(G_i)$ necessarily colours all vertices of e the same. If we perturb this colouring by re-colouring vertex $\ell(e)$ with one of at least $k - \Delta_1$ colours, then we will obtain a colouring in $\Omega_k(G_i)$. However for a given edge and colouring χ in $\Omega_k(G_i)$ there is at most one colouring of $\Omega_k(G_{i-1})$ which could be perturbed to χ.

Thus, since $\Omega_k(G_i) \subseteq \Omega_k(G_{i-1})$ we have

$$
\begin{aligned}
\left(|\Omega_k(G_{i-1})| - |\Omega_k(G_i)|\right)(k - \Delta_1) &\leq |\Omega_k(G_i)| \\
\Rightarrow \quad \frac{1}{2} \leq \frac{k - \Delta_1}{k - \Delta_1 + 1} &\leq \frac{|\Omega_k(G_i)|}{|\Omega_k(G_{i-1})|} \leq 1.
\end{aligned}
$$

We wish to apply the Markov chain Monte Carlo Lemma. Let

$$\varepsilon_1 = \frac{\varepsilon(2 - \varepsilon)}{2(4m + \varepsilon(2 - \varepsilon))}.$$

Let $\xi_i = |\Omega_k(G_i)| / |\Omega_k(G_{i-1})|$. Let Z_i be the indicator random variables that take value 1 if $\mathcal{M}(G_{i+1}, k)$, started from a fixed state and simulated for $T = \tau(\varepsilon_1)$ steps, is a k-colouring of G_{i-1}, and value 0 otherwise. Let $\mu_i = \mathbf{E}(Z_i)$. Since we are simulating for T steps, by Lemma 10 we know that the variation distance of $\mathcal{M}$ from the uniform distribution is no more than ε_1, hence, by Proposition 1, $\xi_i/(1 + \varepsilon(2 - \varepsilon)/4m) \leq \mu_i \leq \xi_i(1 + \varepsilon(2 - \varepsilon)/4m)$.

We will obtain our estimators, $\bar{Z}_i$, for each of the μ_i, by taking the mean of $s = \lceil 165m/\varepsilon^2 \rceil$ independent instances of Z_i (see the Remark following the Markov chain Monte Carlo Lemma).

Applying the Markov chain Monte Carlo Lemma, we have with probability at least $^3/_4$:

$$\frac{1}{1 + \varepsilon} \xi_0 \xi_1 \ldots \xi_{m-1} \leq \bar{Z}_0 \bar{Z}_1 \ldots \bar{Z}_{m-1} \leq (1 + \varepsilon)\xi_0 \xi_1 \ldots \xi_{m-1}.$$

Thus we have an *fpras* for counting the number of k-colourings of a hypergraph that requires a total of msT steps, where $s = \lceil 165m/\varepsilon^2 \rceil$ and $T = \tau(\varepsilon(2-\varepsilon)/(4m + \varepsilon(2 - \varepsilon)))$.

We have therefore illustrated that the algorithm is polynomial in m, n, and ε^{-1}, and thus this randomized approximation scheme is fully polynomial.

4.1.4 Conclusions

To conclude then, we have exhibited an *fpras* for hypergraph k-colourings provided $k > \Delta_1 + \Delta_3$, or $k \geq 2\Delta_2$, and *inter alia* a fully-polynomial almost uniform sampler for hypergraph k-colourings, provided that the same conditions hold on k.

It is natural to inquire as to whether a weaker condition on k will suffice to ensure the existence of an *fpras*. Certainly, no such algorithm can exist for $3 \leq k < \Delta_2$ or $3 \leq k < \Delta_3$, unless $\mathrm{NP} = \mathrm{RP}$, since by a result of Garey, Johnson, and Stockmeyer [52], deciding if such a colouring exists is NP-complete for graphs.

For the case $k = 2$, there is a simple reduction from NOT-ALL-EQUAL-SAT to the hypergraph k-colouring problem: consider a hypergraph with vertex set the set of literals of the NOT-ALL-EQUAL-SAT instance; edges are either a literal and its complement or the set of literals in a clause. Since there are precisely two hypergraph 2-colourings for each satisfying NOT-ALL-EQUAL-SAT assignment, we have that hypergraph 2-colouring is essentially parsimoniously inter-reducible with NOT-ALL-EQUAL-SAT, and thus unless $\mathrm{NP} = \mathrm{RP}$, there can be no *fpras* for hypergraph 2-colourings — even if we restrict the hypergraphs to be 3-uniform (corresponding to NOT-ALL-EQUAL-3SAT[2]). In Section 5.3.3.2 we show that even if we restrict the NOT-ALL-EQUAL-SAT instances to have variables appearing at most twice, counting the number of satisfying assignments is still #P-complete, but in this case the existence of an *fpras* is an open problem. This is equivalent to counting all of the sink- and source-free orientations of a graph, which we consider in Section 5.3. We consider the allied problem of counting sink-free orientations in Sections 4.2 and 5.2.

In Appendix A, we offer an alternative proof of the rapid mixing of the Markov chain $\mathcal{M}(G, k)$, for $k > \Delta_1 + \Delta_3$ using Dobrushin's uniqueness criterion. In Section 5.9 we offer a third, stronger proof of rapid-mixing of $\mathcal{M}(G, k)$, this time for $k \geq \Delta_1 + \Delta_4$.

4.2 Sink-Free Graph Orientations and TWICE-SAT

Graph orientation problems have a long pedigree both in pure mathematics and theoretical computer science. We consider here all of the major combinatorial problems associated with sink-free graph orientations: decision, construction, listing, counting, approximate counting and approximate sampling. We also observe a close connexion between sink-free graph orientations and a restricted form of the SAT problem in which each variable may appear at most twice; we name this TWICE-SAT. We show that this problem is #P-complete, and offer an *fpras* for it. An earlier version of this section appeared in Bubley and Dyer [20]. Following this earlier version Huber [64] has shown, using coupling from the past, how to obtain exact samples from the Markov chain we propose.

4.2.1 Introduction

Counting satisfying assignments to SAT instances is a classic #P-complete problem #SAT [125]. Since SAT is itself NP-complete, it is asking too much even to *approximate* the number of solutions to an arbitrary instance of SAT. For example, the existence of an *fpras* for #SAT would imply $\mathrm{NP} = \mathrm{RP}$. The situation is no better for 2SAT, where

---

[2]That NOT-ALL-EQUAL-3SAT is NP-complete is a result of Schaefer [107].

the decision problem is easy, but the counting problem remains #P-complete. There can be no *fpras* for #2Sat unless NP = RP, even if no variable appears negated. This problem is easily seen to be equivalent to counting all independent sets in a graph. A simple argument [111, Theorem 1.17] rules out the existence of an *fpras* for certain problems — including counting independent sets of all sizes in graphs — unless NP = RP.

We consider other related #Sat restrictions in Appendix B.

Here we consider a different easy restriction of Sat, which we will call Twice-Sat. The restriction is that every Boolean variable should appear at most *twice* altogether in the instance. Showing that the decision problem is in P appears as Problem 9.5.4(a) in [97, p. 207]. We show that the corresponding counting problem #Twice-Sat remains #P-complete, but that in this case an *fpras does* exist. As far as we are aware, this is the first easy case of approximating a natural class of hard instances of #Sat. Our result is the "best possible" in the sense that relaxing the restriction on Sat instances to allow each variable to appear at most *three* times (Thrice-Sat) gives an NP-complete decision problem [97, p. 183] and hence no *fpras* is likely to exist.

There is a close connexion between Twice-Sat and a problem concerning the set of *orientations* of an undirected graph which have no *sink* (vertex with out-degree zero). We call this problem SFO (Sink-Free Orientations). Thus our work relates also to the study of graph orientations, which has been a fruitful source of computational problems. For example, Stanley [115] showed that the number of *acyclic* orientations of a graph is equal in modulus to the value at -1 of its chromatic polynomial. Linial [86] showed that computing this number is a #P-complete problem. Alon, Frieze, and Welsh [6] gave an *fpras* for its dual, totally cyclic orientations in the dense graph case, but the primal problem remains open. Mihail and Winkler [95] showed #P-completeness of, and gave an *fpras* for, counting Eulerian orientations of a graph.

The plan of Section 4.2 is as follows. In Section 4.2.1.1 we describe the problem SFO. In Section 4.2.1.2 we indicate the equivalence between this and Twice-Sat. We show in Section 4.2.2 that decision and construction may be accomplished in linear time. (This gives *inter alia* a complete solution to Problem 9.5.4(a) of [97].) In Section 4.2.3 we show that exact counting is #P-complete. In Section 4.2.4, we develop a fully-polynomial almost uniform sampler and an *fpras*, using the Markov chain Monte Carlo method.

4.2.1.1 *Notation and preliminaries*

Let $G = (V, E)$ be an undirected graph (or multigraph) with $n = |V|$ and $m = |E|$. When we consider algorithmic issues we assume that G is represented as an adjacency-list structure. We use standard terminology. In addition, each edge may be designated *ordinary* or *skew*. An *ordinary* graph is one with no skew edges.

An orientation of an edge e, with end-points u, v is a function $O_e : \{u, v\} \rightarrow \{head, tail\}$. If $O_e(u) = head$, we will say that u is a head of e, otherwise we will say that u is a tail of e. We place a different condition on O_e depending on whether e is ordinary or skew. If e is ordinary, O_e must be a bijection; if e is skew, O_e must be a constant function. Note that in each case there are exactly two possible functions, and hence each edge has two possible orientations. Intuitively, an oriented ordinary edge

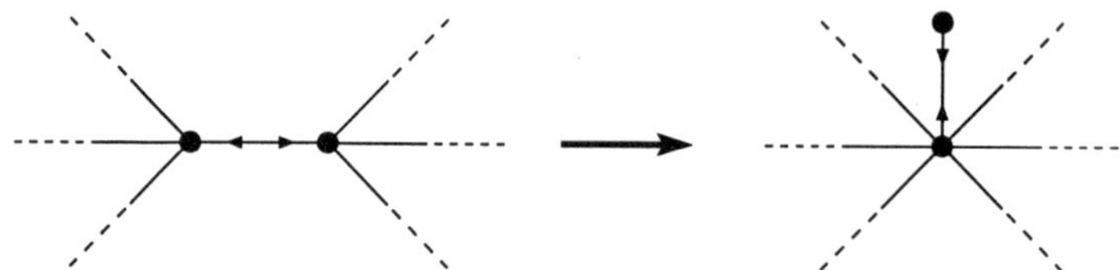

Figure 4.1: A skew-contraction.

"points towards" (is a head of) one end-point and "points away" (is a tail of) the other end-point, whereas an oriented skew edge either points away from both end-points, or points towards both end-points. An *orientation* O of G is an orientation O_e for all $e \in E$. We may *reverse* the orientation of an edge e, by replacing O_e by its unique alternative $\overline{O_e}$, i.e. if an end-point is a head of e it will become a tail and vice versa. If e is reversed in O we will denote the resulting orientation by $\mathrm{rev}(O, e)$. Observe that our definition of orientation is more general than elsewhere in the literature, where a graph orientation normally means that all edges are ordinary.

The *out-degree* (or in-degree) of vertex $v \in V$ in orientation O is the number of edges for which v is the tail (or the head, respectively). Then v is a *sink* (resp. source) of O if its out-degree (or in-degree, respectively) is zero. Orientation O is *sink-free* (or source-free, respectively) if no vertex is a sink (or source, respectively) of O. The decision problem SFO on instance G is then to determine whether G has a sink-free orientation. The derived counting problem #SFO is to determine the number of sink-free orientations of G.

For a graph G, we shall use the notation $\mathcal{N}_i(G)$ to indicate the set of orientations of G with i sinks, and $N_i(G) = |\mathcal{N}_i(G)|$. We shall also use $\mathcal{I}_i(G)$ to denote the set of independent sets of size i in G, and $I_i(G) = |\mathcal{I}_i(G)|$. Where no ambiguity will arise, we abbreviate these to $\mathcal{N}_i$, N_i, $\mathcal{I}_i$, and I_i. For a graph G of which e is an edge, let $G - e$ denote the graph G with edge e *deleted*. If e is not a loop, and is ordinary, then $G \cdot e$ will denote the graph G with edge e *contracted*, in the normal fashion, by deleting e and identifying its end-points. If however e is a skew edge, then $G \cdot e$ will denote the graph G with edge e *skew-contracted* (see Figure 4.1). In this case the end-points of e are identified, but e itself is not deleted. An additional vertex is created, so that e now joins the new vertex to that resulting from the identification, i.e. e becomes an isthmus.

The deletion of an edge in a sink-free orientation induces a unique orientation in the resulting graph which may not be sink-free. However, the contraction of an ordinary edge in a sink-free orientation always induces a sink-free orientation of the resulting graph. To preserve this property in the case of skew-contraction, we insist that e is oriented to point away from its end-points in the resulting graph. Thus a skew-contraction also induces a unique sink-free orientation.

4.2.1.2 Equivalence of Twice-SAT and SFO

Suppose we have a TWICE-SAT instance with m variables, $x_1, x_2, \ldots, x_m$, and n clauses, $C_1, C_2, \ldots, C_n$. Consider now the graph G with vertices $C_1, C_2, \ldots, C_n$, and edges $x_1, x_2, \ldots, x_m$. If x_i appears only once in the instance, we treat it as if it appears twice in the same clause. The edge x_i will have as end-points the two (not necessarily distinct) clauses in which x_i appears — if the two appearances are as the same literal, then the edge is skew; if as a literal and its negation, then the edge is ordinary. We identify one truth value of each variable with one orientation of the corresponding edge, and the other with its reversal in an obvious fashion. Then solving SFO on G solves the TWICE-SAT instance. Clearly this reduction is parsimonious, as is the reverse transformation.

4.2.2 Decision and Construction

In this section we show that the problems of deciding SFO, and constructing a witness if one exists, are computationally simple tasks. Specifically, for a graph with n vertices and m edges, we may accomplish the tasks of construction and decision in time $O(m + n)$. To show this, we first provide a Lemma characterizing graphs which have a sink-free orientation.

LEMMA 11. *A connected graph, G, has a sink-free orientation if it is either ordinary and has a cycle or it has a skew edge.*

Proof. Clearly if G has no cycles it has more vertices than edges. Then it is immediate that no orientation can be sink-free if G is ordinary, since each vertex may be mapped injectively to an incident edge pointing away from it in a sink-free orientation.

If G has a cycle, we may construct a sink-free orientation. Find a spanning tree, T, of G. Let v be a vertex of G which is incident on edge $e \notin T$. Such a vertex must exist. Orient the ordinary edges of T so that they "point towards" v and the skew edges to point away from both end-points. Finally orient e to point away from v. Any remaining edges may be oriented in arbitrary fashion.

It remains only to show that a tree with at least one skew edge has a sink-free orientation. Again we orient all skew edges away from their end-points, and orient ordinary edges to point towards any skew edge. $\qquad\square$

COROLLARY 3. *A graph G has a sink-free orientation if and only if no component of G is an ordinary tree[3].*

Proof. This follows from the fact that the number of sink-free orientations of a graph is the product of the number of sink-free orientations of its components. $\qquad\square$

COROLLARY 4. *Deciding SFO, and constructing a witness if one exists, may be accomplished in time linear in the number of vertices and edges of the graph.*

Proof. The method explained in the proof of Lemma 11, using Corollary 3 may readily be implemented by breadth- or depth-first search. $\qquad\square$

The reader may wonder why we do not simply use the skew edges to reduce the problem to a smaller ordinary graph. Though this is clearly valid for decision, *it is not so for counting*, which is our eventual aim.

---

[3]In this context, an isolated vertex is a tree.

4.2.3 Exact Counting

4.2.3.1 Self-reducibility

The deletion of an edge in a sink-free orientation induces a unique orientation in the resulting graph which may not be sink-free. However, the contraction of an ordinary edge in a sink-free orientation always induces a unique sink-free orientation of the resulting graph, and recall that we have defined the contraction of a skew edge (a skew contraction) to preserve this property.

LEMMA 12. *Suppose $G = (V, E)$ is a graph, and $e \in E$ is a non-loop edge. Then there is a natural bijection between $\mathcal{N}_0(G)$ and $\mathcal{N}_0(G - e) \cup \mathcal{N}_0(G \cdot e)$.*

Proof. Let $<$ denote an arbitrary total order on V. We will denote the end-points of e by a and b, and without loss of generality, assume $a < b$. Define $F : \mathcal{N}_0(G) \to \mathcal{N}_0(G - e) \cup \mathcal{N}_0(G \cdot e)$ in the following way: if O is a sink-free orientation of G in which e points towards a, and the orientation obtained from O by reversing the orientation of e is also a sink-free orientation, then $F(O) = O - e \in \mathcal{N}_0(G - e)$. Otherwise, $F(O) = O \cdot e \in \mathcal{N}_0(G \cdot e)$. To see that this function does generate orientations in the range claimed, consider the two cases used in the construction of F — in the first case, there must exist edges oriented away from both a and b in O, hence $O - e \in \mathcal{N}_0(G - e)$; in the second case there must exist at least one edge oriented away from a or b if e is ordinary; if e is skew, then the definition of a skew-contraction ensures that the contracted vertex is not a sink, hence $O \cdot e \in \mathcal{N}_0(G \cdot e)$. Furthermore, F is clearly injective, and a moment's thought establishes that it is surjective too. $\square$

We remark that this bijection may be generalized to orientations with i sinks.

COROLLARY 5. $N_0(G) = N_0(G - e) + N_0(G \cdot e)$ *and* $N_0(G \cdot e) \geq N_0(G - e)$.

Proof. The first claim is obvious. The second follows from examining the map F in the proof of Lemma 12. $\square$

COROLLARY 6. *SFO is a self-reducible decision problem.*

Proof. This is immediate from Lemma 12 and the definition of self-reducibility [108]. $\square$

COROLLARY 7. *Listing all sink-free orientations of a graph has a polynomial space, polynomial delay [72] algorithm.*

Proof. We may use the recurrence of Lemma 12 as the basis of a recursive algorithm which has as base cases those graphs which have no sink-free orientations (which we know from Corollary 3) and graphs in which all edges are either loops, or skew edges incident on a vertex of unit degree. The base cases are elementary. Provided we ensure that the edge chosen for the recursion is never a loop or a skew edge incident on a vertex of unit degree, and that the recursion is evaluated lazily, this algorithm is clearly polynomial space. To see that it has polynomial delay, consider the recursion tree as a search tree in the algorithm. Appealing once more to Corollary 3 we see that this search tree will be efficiently pruned by the algorithm, and thus in traversing from

Figure 4.2: Two graphs with identical Tutte polynomials.

one sink-free orientation to the next, we are making $O(m)$ moves on the search tree, which completes the proof. $\square$

Remark. Valiant [125] notes the existence of a similar recursive argument, which, as Goldberg [55] points out, extends to any self-reducible structure. $\bigcirc$

The contraction-deletion formula of Lemma 12, and that defining the *Tutte polynomial* (see Section 1.4.1) are strikingly similar. However, SFO does not represent any point in the "Tutte plane". Indeed we can prove a stronger statement than this:

PROPOSITION 6. *The number of sink-free orientations of a graph is not a function of the Tutte polynomial of the graph.*

Proof. The standard recursive definition of the Tutte polynomial (see Section 1.4.1) defines, for a loop e, the following reduction: $T(G; x, y) = yT(G - e; x, y)$.

It is clear from this reduction that the two graphs in Figure 4.2 share the same Tutte polynomial; however, by inspection, it can be seen that the graph on the left admits 32 sink-free orientations, whereas the graph on the right admits only 24. $\square$

4.2.3.2 Relationship to counting independent sets

Denote the degree!sequence of graph G by $d_1, d_2, \ldots, d_n$ for vertices $v_1, v_2, \ldots, v_n$. We will use $\mathrm{Stab}(G)$ to denote the stability number of G: the size of the largest independent set.

The following discussion relates only to *connected ordinary* graphs. Note that, in an orientation with i sinks, the sinks form an independent set in G. Suppose we consider the following process for generating an orientation with at least j sinks. Choose an independent set $\{v_{\lambda_1}, v_{\lambda_2}, \ldots, v_{\lambda_j}\} \in \mathcal{I}_j$ and orient the edges incident on these to make them sinks. Then orient the other edges arbitrarily. This gives $2^{m-d_{\lambda_1}-d_{\lambda_2}-\cdots-d_{\lambda_j}}$ distinct orientations. Summing this over all independent sets clearly counts all orientations with at least j sinks, but those with $i \geq j$ sinks will be counted $\binom{i}{j}$ times, i.e.

$$\sum_{i=j}^{\mathrm{Stab}(G)} \binom{i}{j} N_i = \sum_{\{v_{\lambda_1}, v_{\lambda_2}, \ldots, v_{\lambda_j}\} \in \mathcal{I}_j} 2^{m-d_{\lambda_1}-d_{\lambda_2}-\cdots-d_{\lambda_j}}. \tag{4.6}$$

We may invert this to give

$$N_k = \sum_{j=k}^{\mathrm{Stab}(G)} (-1)^{j+k} \binom{j}{k} \sum_{\{v_{\lambda_1}, v_{\lambda_2}, \ldots, v_{\lambda_j}\} \in \mathcal{I}_j} 2^{m-d_{\lambda_1}-d_{\lambda_2}-\cdots-d_{\lambda_j}}. \tag{4.7}$$

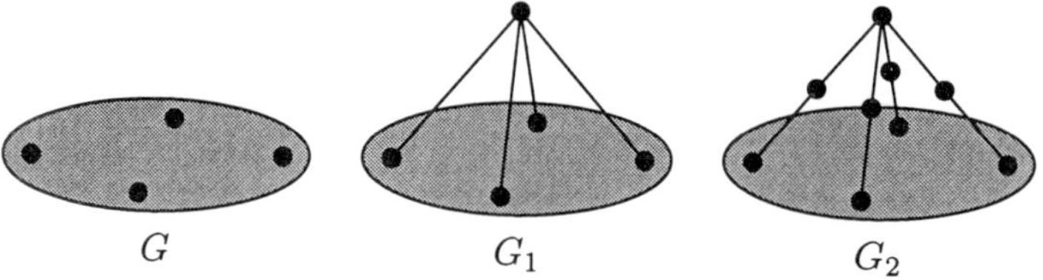

Figure 4.3: Constructing graphs G_i.

From equation (4.7) we can, for example, deduce a formula for the number of sink-free orientations of the complete ordinary graph on r vertices, K_r, i.e.

$$N_0(K_r) = 2^m - r2^{m-r+1},$$

where $m = r(r-1)/2$.

Note that if G is regular of degree Δ, then equation (4.6) reduces to:

$$\sum_{i=k}^{n} \binom{i}{k} N_i = 2^{\Delta(n-2k)/2} I_k. \tag{4.8}$$

We will use this below.

4.2.3.3 #P-completeness of #SFO

We show that counting version of SFO is a #P-complete problem.

THEOREM 3. *Counting the number of sink-free orientations of a simple ordinary graph is #P-complete.*

Proof. Let P_i be the path graph on $(i+1)$ vertices. Note that P_i has exactly $(i+1)$ orientations which make no internal vertex into a sink, and exactly one of these makes a particular end vertex into a sink. For a connected graph G on n vertices with m edges, define new graphs G_i for each natural number i in the following manner (see Figure 4.3): form the union of G, and n copies of P_i in such a fashion that the vertices of the n copies of P_i are distinct from each other and from those of G, with the following exceptions: each copy of P_i shares exactly one vertex, say w with the others, and w is an end-point of each of the P_i; the other end-point is a vertex of G.

If O is an orientation of G with k sinks, then this may be extended to a sink-free orientation of G_i in $(i+1)^{n-k} - i^{n-k}$ ways: the P_i incident on a sink of G can be oriented in only one way; the remaining P_i can be oriented in any of $(i+1)$ ways, but not all can be oriented in the i ways which point into w. Hence

$$N_0(G_i) = \sum_{k=0}^{n-1} N_k(G) \left[(i+1)^{n-k} - i^{n-k} \right].$$

Since the total number of orientations of a graph is 2^m, we have

$$
\begin{pmatrix}
1 & 1 & 1 & \cdots & 1 \\
1 & 2 & 2^2 & \cdots & 2^{n-1} \\
\vdots & \vdots & \vdots & \ddots & \vdots \\
1 & n-1 & (n-1)^2 & \cdots & (n-1)^{n-1}
\end{pmatrix}
\times
\begin{pmatrix}
N_{n-1}(G) \\
N_{n-2}(G) \\
\vdots \\
N_0(G)
\end{pmatrix}
=
$$

$$
\begin{pmatrix}
2^m \\
N_0(G_1) + 2^m \\
\vdots \\
N_0(G_{n-1}) + \cdots + N_0(G_1) + 2^m
\end{pmatrix}.
$$

Observe that the matrix on the left is a Vandermonde matrix with distinct columns, and hence it is non-singular. Since matrix inversion and multiplication may be accomplished in polynomial time, we have therefore that the task of calculating $N_i(G)$ for all $i > 0$ is no harder (up to a polynomial factor) than calculating $N_0(G)$, the number of sink-free orientations.

Now suppose that $G = (V, E)$ is a regular connected graph of degree Δ. Then equation (4.8) implies that if we can calculate the $N_i(G)$, we can calculate the $I_i(G)$.

It is unclear whether one may adapt an NP-completeness proof of Garey, Johnson, and Stockmeyer [52, Theorem 2.6] to show that this directly implies #P-completeness. A result of Vadhan [123] shows that this computation is #P-complete if G is a regular graph of degree 5. This may be tightened slightly.

If H is a graph, and $L(H)$ its line-graph, then the independent sets of size k in $L(H)$ are in 1–1 correspondence with the matchings of size k in H. Observe further that if H is regular of degree 3, then $L(H)$ is regular of degree 4. Hence we may calculate the number of perfect matchings in connected regular graphs of degree 3, from which it follows that we may calculate the number of perfect matchings in regular graphs of degree 3. But by a result of Dagum and Luby [33, Theorem 6.2] this is #P-complete. Since #SFO is clearly in #P, the result is established. $\square$

COROLLARY 8. *Counting the number of satisfying assignments to* SAT *instances in which each literal appears exactly once is #P-complete.*

Proof. This follows immediately from our discussion in Section 4.2.1.2 about the equivalence of Twice-SAT and SFO. $\square$

If all edges are skew, the problem becomes a counting version of the problem EDGE COVER [51], counting covers of all sizes. Note that our results below thus illustrate an *fpras* for #EDGE COVER. Note that there is in fact a simpler #P-completeness proof in this case. By constructing a sequence of graphs, G'_k, where G'_k is the graph formed by swapping every edge of G for a multi-edge of multiplicity k, we may use a similar method to that used in the proof of Theorem 3 to calculate the number of covers of each size. The number of covers of size $n/2$ is the number of perfect matchings. We consider additional restrictions of the #EDGE COVER problem in Appendix B.

4.2.4 Approximate Sampling

In this section we provide a fully-polynomial!almost uniform sampler, and a fully-polynomial randomized approximation scheme for sink-free orientations of a graph, and hence for TWICE-SAT.

We have from Corollary 6 that SFO is self-reducible, and thus (although we make no use of this here) theorems of Jerrum, Valiant, and Vazirani [71] and Sinclair and Jerrum [112] apply, which state that a fully-polynomial almost uniform sampler for a self-reducible problem exists if and only if a fully-polynomial randomized approximation scheme does.

We note that a facile algorithm for sampling and approximation exists: by choosing random orientations of all edges and discarding resulting graph orientations with sinks, we have a (uniform) sampler for sink-free orientations. By repeatedly choosing random orientations, we may use a Monte Carlo method to estimate the number of sink-free orientations. If p is the proportion of orientations that are sink-free, the sampler will run in expected time $O(mp^{-1})$ and the Monte Carlo estimation method will yield a randomized approximation scheme with error parameter ε that runs in time $O(mp^{-1}\varepsilon^{-2})$. Let δ be the minimum degree in G, and S the number of sinks in O. Then, for a random orientation O,

$$(1 - p) = \mathbf{P}(S > 0) \leq \mathbf{E}(S) = \sum_{i=1}^{n} 2^{-d_i} \leq m2^{-\delta}.$$

Thus a sufficient condition for the sampler and the randomized approximation scheme to be fully polynomial is to have $\delta > \lceil \log_2 m \rceil$. This deals with the case of "relatively dense" graphs, but for sparse graphs this procedure may clearly fail. For example, consider a regular bipartite graph G of degree Δ. This has an independent set of size $n/2$, and hence

$$\mathbf{P}(S = 0) \leq \left(1 - 2^{-\Delta}\right)^{n/2}.$$

Thus for, say, $\Delta \leq \frac{1}{2}\log_2 n$, the probability that sampling random orientations will succeed is exponentially small.

However, we will exhibit a fully-polynomial almost uniform sampler by defining a Markov chain with state space the set of sink-free orientations of a graph, and proving that it is rapidly mixing. By Corollary 4 we may quickly find an initial sink-free orientation with which to start the simulation.

The technique that we shall use to prove that the Markov chain is rapidly mixing is *coupling*.

Throughout the rest of this section we will make three simplifying assumptions: firstly that our input graph is connected, secondly that it contains no vertex of unit degree, and thirdly that it is not an ordinary cycle. To justify these assumptions, observe that to sample from disconnected graphs, it suffices to sample independently from each of the component graphs. Suppose the graph has a vertex of unit degree, and its adjacent edge is ordinary. In any sink-free orientation then, this edge must be oriented towards the vertex of unit degree, and we may thus delete the vertex and edge without materially affecting the sampling. If the graph has a vertex of unit degree, and its adjacent edge is skew, then we may replace the skew edge and its end-point of unit degree by an ordinary loop, again without materially affecting the sampling. If the graph is an ordinary cycle, it has just two sink-free orientations from which we may trivially sample.

In a sink-free orientation, we will call a vertex *critical* if it has unit out-degree; we will call an edge *critical* if it is oriented away from a critical vertex. Our assumptions on G ensure that there is at least one non-critical edge in any sink-free orientation of G. To see this, note that the number of critical edges is at most $(n - s)$, where s is the number of skew edges oriented away from their end-points. If G has more than n edges, we are done. But otherwise G must be a cycle containing at least one skew edge. But then all skew edges cannot be oriented towards their end-points, since in this case no sink-free orientation is possible.

For a graph $G = (V, E)$ consider the Markov chain, $\mathcal{M} = \mathcal{M}(G)$, with time states (X_t) which take values in the state space $N_0(G)$ (the sink-free orientations). The transition probabilities for $\mathcal{M}(G)$ are modelled by: with probability $1/2$ choose $e \in E$ uniformly at random. If e is not critical then reverse its orientation. Otherwise remain in the same state.

PROPOSITION 7. *Provided G either has more than one cycle or has a skew edge, $\mathcal{M}(G)$ has uniform stationary distribution.*

Proof. The fact that $\mathcal{M}(G)$ is irreducible will follow from the coupling argument below. We show that random walks started in any two states can be made to meet in finite time with non-zero probability. This is sufficient to establish irreducibility. $\mathcal{M}(G)$ is trivially aperiodic, from the "do nothing" conditions. Since $\mathcal{M}(G, k)$ is irreducible and aperiodic, it is ergodic, and, from the detailed balance equation (1.2), it follows that its stationary distribution is uniform. $\qquad\square$

We wish to construct a joint process (X, Y), such that marginally X and Y are faithful copies of $\mathcal{M}$, and X and Y have a probabilistic tendency to couple. We will define this process by:

1. Choose an edge e uniformly at random.

2.
   - If e is critical in both X and Y:
     Do nothing.
   - If e is critical in X but not Y:
     With probability $1/2$ reverse the orientation of e in Y.
   - If e is critical in Y but not X:
     With probability $1/2$ reverse the orientation of e in X.
   - If e is critical in neither X nor Y:
     - If e has the same orientation in both X and Y:
       With probability $1/2$ reverse the orientation of e in X and Y.
     - If e has a different orientation in X and Y:
       With probability $1/2$ reverse the orientation of e in X, otherwise reverse the orientation of e in Y.

Let A_t be the set of edges that agree in orientation between X_t and Y_t, and let $\bar{A}_t$ be its complement. (We will write A, $\bar{A}$ where the context is clear.) Let C_X and C_Y be the set of critical edges in X and Y respectively.

Define $D_1 = D_1(X_t, Y_t) = |\bar{A}_t|$; this is the Hamming distance between X and Y. Let D_2 be the minimum number of transitions to go from the current state of the joint process to one at which some transition could decrease D_1.

PROPOSITION 8. *At any state of the joint process, the probability that D_1 increases at the next transition is bounded above by the probability that D_1 decreases. When $D_2 = 0$, the probability that D_1 decreases is at least $1/2m$.*

Proof. From the definition of the coupling, it is easy to see that D_1 will only increase if the chosen edge e is in $A \cap (C_X \oplus C_Y)$ (where $\oplus$ is disjoint union), and we make a non-null transition. This has probability $|A \cap (C_X \oplus C_Y)|/2m$. On the other hand, D_1 will decrease only if the edge we choose is in $\bar{A} \setminus (C_X \cap C_Y)$, and we make a non-null transition. If $e \in \bar{A} \setminus (C_X \cup C_Y)$, then D_1 will decrease by 1 with certainty, but if $e \in \bar{A} \cap (C_X \oplus C_Y)$ it will decrease only with probability $1/2$. Thus the probability of a decrease is $|\bar{A} \setminus (C_X \cup C_Y)|/m + |\bar{A} \cap (C_X \oplus C_Y)|/2m$. Therefore, to prove the first part of the proposition, it suffices to show that

$$2|\bar{A} \setminus (C_X \cup C_Y)| + |\bar{A} \cap (C_X \oplus C_Y)| \geq |A \cap (C_X \oplus C_Y)|.$$

To do this, we will construct an injective mapping from $A \cap (C_X \oplus C_Y)$ to $[(\bar{A} \setminus (C_X \cup C_Y)) \times \{0, 1\}] \cup [\bar{A} \cap (C_X \oplus C_Y)]$.

First note a couple of points. There are at least two edges incident upon each vertex, from our assumptions on G. Also there can be at most one edge in $A \cap (C_X \oplus C_Y)$ oriented away from any vertex v, since in one of X or Y, all other edges are oriented into v. For $e \in A \cap (C_X \oplus C_Y)$, oriented away from v, there must be at least one other edge, $e' \in \bar{A}$ incident on v, since e is critical in only one of X and Y. Clearly e' is not critical at v, although it may be critical at its other end-point, w say.

If e' is critical at w in either X or Y, then we map e to e', since $e' \in \bar{A} \cap (C_X \oplus C_Y)$. Since e' is critical at w, all other edges incident on w must be oriented towards it in either X or Y. Thus there can be no edge in $A \cap (C_X \oplus C_Y)$ oriented away from w. Hence we are assured that this portion of the map is injective.

Therefore suppose $e' \in \bar{A} \setminus (C_X \cup C_Y)$. Let $<$ be a total order on V, and let $r = 0$ if $v < w$, $r = 1$ otherwise. Now we will map e to (e', r). We have already established that there is at most one edge in $A \cap (C_X \oplus C_Y)$ oriented away from either v or w, and hence the map is injective.

The final claim is evident from the definitions of the random walk and D_2. $\qquad\square$

PROPOSITION 9. *If $D_2 > 0$ then all edges in $\bar{A}$ are critical in both X and Y (and therefore not skew) and the edges in $\bar{A}$ form vertex-disjoint (directed) cycles.*

Proof. If $D_2 > 0$ then every edge in $\bar{A}$ must be critical in both X and Y, since otherwise we could decrease D_1. Edges that are critical in both X and Y and have different orientation can clearly not be skew.

If a vertex v is incident on an edge $e \in \bar{A}$, then all edges in A incident on v must be oriented towards v, for the criticality of e would otherwise be violated in either X

or Y. A vertex incident on a loop $e \in \bar{A}$ cannot be incident on any other edge in $\bar{A}$, since otherwise e would not be critical in both X and Y. Furthermore, a vertex incident on a (non-loop) edge $e \in \bar{A}$ must be incident on precisely one other non-loop edge $\bar{A}$. If it were incident on no such edge, then the edge could not be critical in both X and Y and satisfy the sink-free property. If it were incident on two or more such edges in $\bar{A}$, the vertex could not be critical in both X and Y. Hence the edges in $\bar{A}$ form vertex-disjoint directed cycles. $\qquad\square$

LEMMA 13. *If $D_2 > 0$, then $(D_2 - 1)$ is the shortest path distance in G from an edge in $\bar{A}$ to a non-critical edge.*

Proof. Note that there is a non-critical edge, so the path distance is well-defined. We will use induction. The claim is true for $D_2 = 1$, since then there must be a non-critical edge, adjacent to an edge in $\bar{A}$, whose reversal will render an edge in $\bar{A}$ non-critical. Suppose therefore that the claim is true for $D_2 < r$, and we have an orientation O with path distance r. Let e be a closest non-critical edge to $\bar{A}$. Reversing e will make some edge at path distance $(r - 1)$ non-critical, and hence the path distance will be at most $(r - 1)$. Hence $D_2 \le (r - 1)$ in $\mathrm{rev}(O, e)$. Thus $D_2 \le r$ in O. But we cannot have $D_2 < r$ in O, or the inductive hypothesis is contradicted. $\qquad\square$

PROPOSITION 10. *Whatever the state of the joint process, the probability of D_2 increasing at the next transition is either zero or $1/2m$. Furthermore, if $D_2 > 0$ then the probability that D_2 decreases is bounded below by $1/2m$.*

Proof. Using Lemma 13, it follows that D_2 can increase only in the following circumstances. There is a unique path of length D_2 from a vertex u, incident on $\bar{A}$, to a vertex w. The edge $e = \{v, w\}$ is non-critical, the edges between u and v (if any) are all critical. However, the reversal of e makes it critical. This proves the first claim, since e is unique if it exists. For the second, note that the reversal of the non-critical edge at the end of any shortest path (from $\bar{A}$ to a non-critical edge) will reduce D_2. There is always at least one such edge. $\qquad\square$

COROLLARY 9. *If $D_2 > 0$, $D_1 + D_2 - 1 \le |C_X \cap C_Y| \le n$.*

Proof. Note that $D_1 = |\bar{A} \cap (C_X \cap C_Y)|$ from Proposition 9. From Lemma 13, it follows that $(D_2 - 1) \le |A \cap (C_X \cap C_Y)|$. Adding, we have the first inequality. The second follows from the fact that there is at least one critical vertex for each critical edge. $\qquad\square$

Whilst we will not use the next corollary here, we will use it in Section 5.2 when we revisit the problem of Twice-Sat.

COROLLARY 10. *The diameter of $\mathcal{M}$ is bounded above by $\max\{2n + 1, m\}$.*

Proof. Given two states X and Y, we see that in order to couple them, we will in the worst case have first to decrease D_2 to zero, and then decrease D_1 to zero. Since $\mathcal{M}$ is a reversible Markov chain, we see that we have implicitly constructed a transition path between an arbitrary pair of states. Since change in D_1 results from only one of the pair of coupled states actually performing a non-null transition, whereas change in D_2 results from both of the pair performing transitions, we have that the length of this transition path is $D_1 + 2D_2$. Thus the diameter of $\mathcal{M}$ is bounded above by $\max_{0 \leq d_1 \leq m}\{d_1 + 2\max\{n - d_1 + 1, 0\}\}$ (using Corollary 9), from which the result follows. $\square$

Consider the random process (D_1, D_2) defined on the state space

$$\mathcal{D} = (0,0) \cup \{(d_1, d_2) \mid 1 \leq d_1 \leq m; 0 \leq d_2 \leq \max\{n - d_1 + 1, 0\}\}.$$

Two states are adjacent in $\mathcal{D}$ if and only if their D_2 values are 0 and their D_1 values differ by 1, or their D_1 values are equal and their D_2 values differ by 1. The time to couple is the time for this process to reach the state $(0,0)$. But, from Propositions 8 and 10, it follows that this time is bounded above by the time required for the Markov chain $\mathcal{B}$ on $\mathcal{D}$, with transition probabilities $1/2m$ between adjacent states, to reach $(0,0)$ when started at the state furthest from $(0,0)$, i.e. $(m,0)$. We analyse $\mathcal{B}$ below.

Let E_a be the expected number of steps $\mathcal{B}$ will make before it reaches $(D_1, 0)$, given that it starts at (D_1, a), for $1 \leq D_1 \leq n$. Let $r = (n - D_1 + 1)$, then using conditional expectations and the above, we have

1. $E_0 = 0$,

2. $2E_a = 2m + E_{a-1} + E_{a+1}$ for $0 < a < r$,

3. $E_r = 2m + E_{r-1}$,

where $r = n - D_1 + 1$. Solving by standard techniques, we establish that $E_a = m(a + 2ra - a^2)$, and in particular $E_1 = 2m(n - D_1 + 1)$.

Now let $f(b)$ be the expected time in $\mathcal{B}$ to reach $(0,0)$ given that we start in the state $(b,0)$, $0 \leq b \leq m$.

1. $f(0) = 0$,

2. $f(b) = 1 + f(b-1)/2m + f(b+1)/2m + (E_1 + f(b))/2m + (1 - 3/2m)f(b)$, for $0 < b \leq n$,

3. $f(b) = 1 + f(b-1)/2m + f(b+1)/2m + (1 - 1/m)f(b)$, for $n < b < m$,

4. $f(m) = 1 + f(m-1)/2m + (1 - 1/2m)f(m)$,

where $E_1 = 2m(n - b + 1)$. Rearranging, we have

$$f(b+1) - 2f(b) + f(b-1) = \begin{cases} -2m(n - b + 2) & \text{for } 0 < b \leq n, \\ -2m & \text{for } n < b < m, \end{cases}$$

with boundary conditions $f(0) = 0$, and $f(m) = f(m-1) + 2m$. This has solution

$$f(b) = \begin{cases} bm\left(6m + b^2 - 3bm - 6b - 3n^2 + 6n + 5\right)/3 & \text{for } 0 \leq b \leq n, \\ m\left(6mb + 3b - 3b^2 + n^3 + 3n^2 + 2n\right)/3 & \text{for } n < b \leq m, \end{cases}$$

In particular, we have that $f(m) = m(3m^2 + 3m + n^3 + 3n^2 + 2n)/3$ — clearly $f(m)$ is an upper bound on the expected time for the joint process to couple.

LEMMA 14. *The mixing time of the Markov chain, $\mathcal{M}$, satisfies $\tau \leq \lceil em(3m^2 + 3m + n^3 - n)/3 - 1 \rceil$, and the mixing rate $\tau(\varepsilon) \leq \lceil \ln(\varepsilon^{-1}) \rceil \tau$.*

Proof. Using Markov's inequality, we see that the probability that we have not coupled by time t is bounded above by $f(m)/(t+1)$. In particular, in order to ensure that the probability that we have not coupled is no greater than e^{-1}, it suffices to simulate $\lceil em(3m^2 + 3m + n^3 - n)/3 - 1 \rceil$ steps of the random walk. Since we may run successive, independent coupling "trials" of this length, to ensure that the probability that we have not coupled is no greater than ε we need simulate no more than $\lceil \ln(\varepsilon^{-1}) \rceil \tau$. Finally, an application of the Coupling Lemma completes the proof. $\square$

Thus we have an almost uniform sampler for SFO that runs in time $O(m(m^2 + n^3) \ln \varepsilon^{-1})$, i.e. it is a fully-polynomial almost uniform sampler .

4.2.5 The Approximation Scheme

To avoid trivialities, we will assume that $m > n > 2$, $0 \leq \varepsilon \leq 1$, and that G is connected.

Let $G_n = G$, and for each $1 \leq a < n$ define G_a to be the graph obtained by (skew) contracting an edge of G_{a+1}. This edge should be chosen so that it is neither a loop, nor a skew edge that was created by a skew-contraction. Then

$$N_0(G) = N_0(G_n) = \frac{N_0(G_n)}{N_0(G_{n-1})} \times \frac{N_0(G_{n-1})}{N_0(G_{n-2})} \times \cdots \times \frac{N_0(G_2)}{N_0(G_1)} \times N_0(G_1). \quad (4.9)$$

If a is the number of loops of G_1, then $N_0(G_1) = 2^a$, if G_1 has an edge that is not a skew loop, and $N_0(G_1) = 2^a - 1$ otherwise.

We wish to apply the Markov chain Monte Carlo Lemma. Let

$$\varepsilon_1 = \frac{\varepsilon(2 - \varepsilon)}{2(4(n-1) + \varepsilon(2 - \varepsilon))}.$$

Let $\xi_i = N_0(G_i)/N_0(G_{i+1})$. Recall that in the proof of Lemma 12 we defined a bijection, F, between $\mathcal{N}_0(G)$ and $\mathcal{N}_0(G - e) \cup \mathcal{N}_0(G \cdot e)$ for some non-loop edge e. Let Z_i be the indicator random variable that takes value 1, if $\mathcal{M}(G_{i+1})$, started from a fixed state, and simulated for $T = \tau(\varepsilon_1)$ steps, is an orientation that maps to $\mathcal{N}_0(G \cdot e)$ under F, and value 0, otherwise. Let $\mu_i = \mathbf{E}(Z_i)$. Since we are simulating for T steps, by Lemma 14 we know that the variation distance of $\mathcal{M}$ from the uniform distribution is no more than ε_1, hence, by Proposition 1, $\xi_i/(1 + \varepsilon(2-\varepsilon)/4(n-1)) \leq \mu_i \leq \xi_i(1 + \varepsilon(2 - \varepsilon)/4(n - 1))$.

Since $N_0(G \cdot e) \geq N_0(G - e)$ from Corollary 5, we have that $1/2 \leq \xi_i \leq 1$.

We will obtain our estimators, $\bar{Z}_i$, for each of the μ_i, by taking the mean of $s = \lceil 154(n - 1)\varepsilon^{-2} \rceil$ independent instances of Z_i (see the Remark following the Markov chain Monte Carlo Lemma).

We may thus apply the Markov chain Monte Carlo Lemma, to establish that

$$\frac{1}{1+\varepsilon}\xi_1\xi_2\ldots\xi_{n-1} \leq \bar{Z}_1\bar{Z}_2\ldots\bar{Z}_{n-1} \leq (1+\varepsilon)\xi_1\xi_2\ldots\xi_{n-1},$$

with probability at least $3/4$.

Thus we have an *fpras* for counting sink-free orientations that requires a total of $(n - 1)sT$ steps, i.e. the number of simulated Markov chain steps is $O((n^2m^3 + n^5m)\varepsilon^{-2}\ln(n/\varepsilon))$.

4.2.6 Conclusions

To conclude then, we have classified all of the major combinatorial problems associated with the set of sink-free orientations of a graph, and the set of satisfying assignments to a Boolean formula in conjunctive normal form in which each variable appears at most twice. Decision and construction are essentially linear time, listing is polynomial space, polynomial delay, counting is #P-complete, and there are fully-polynomial algorithms for approximate counting and sampling.

Perhaps the most significant of these results is that there are natural hard cases of #SAT which may be approximated quickly. It is worth noting that, in terms of the SAT problem, the Markov chain we use, and show is rapidly mixing, has an exceptionally simple form — the Markov chain may be described by: with probability $1/2$ pick a variable uniformly at random, and if the result of negating its truth value would be a satisfying assignment, negate its truth value, otherwise do nothing.

4.3 Log-Concave Sampling, and the Volume of a Convex Body

In this section we describe a new method for proving the polynomial-time convergence of an algorithm for sampling (almost) uniformly at random from a convex body in high dimension. Previous approaches have been based on estimating conductance via isoperimetric inequalities. We show that a more elementary coupling argument can be used to give a similar result. A version of this section has appeared as Bubley, Dyer, and Jerrum [26].

4.3.1 Introduction

Dyer, Frieze, and Kannan [42] gave the first fully-polynomial randomized approximation scheme for approximating the volume of a convex body in $\mathbb{R}^n$, for large n. The algorithm was based on a reduction to sampling uniformly at random in such a body. The necessary sampling was achieved, to close enough approximation, by simulating a polynomial number of steps of a random walk in the body.

The random walk employed in [42] was akin to a traditional unbiased random walk on a (sufficiently fine) n-dimensional lattice, but restricted to the interior of the body. The time complexity of the resulting sampling procedure—a high-degree polynomial in the dimension n—has since been dramatically improved in a sequence of papers [8, 40, 50, 88, 74, 75, 87, 49]. (See [73] for a recent survey.) These papers introduced important algorithmic advances, some of which will be exploited here. Specifically, we adopt the idea of using a surrogate log-concave distribution [8], and the use of a "Metropolized" rotationally symmetric random walk [88]. (This random walk is sometimes referred to as the "ball walk" in the literature.)

In parallel with the algorithmic improvements came improvements in analytical technique. The approach of [42] to proving rapid-mixing was to estimate its conductance. In [42], the conductance was bounded by considering the geometric structure of the state space of the random walk in the light of a classical isoperimetric inequality. Subsequent papers introduced new isoperimetric inequalities that were motivated by the application, but still interesting in their own right. The conductance approach has intuitive appeal, provided one treats the geometric results as given; however, the isoperimetric inequalities are quite difficult to prove.

Here we explore this problem from the perspective of *coupling*, which previously seemed difficult to apply to this problem. Specialized to the current situation, the idea, in essence, is to construct two copies of the ball walk (or something very close to it) on a joint sample space in such a way that the two walks converge in a modest number of steps.

By viewing the problem from the correct perspective, we show that it is in fact rather easy to apply the coupling argument to verify polynomial convergence. To achieve this, we first transform the uniform sampling problem to that of generating a random point from the log-concave distribution of [88]. Then we apply our coupling arguments to a suitably chosen random walk for the transformed problem. Our treatment has the advantages of being self-contained and elementary, though we have not succeeded in completely eliminating technical complexities.

The most important application of the sampling algorithm is to the volume estimation problem mentioned at the outset. We do not consider this here, but refer the interested reader to the papers already cited. The best time bound currently known is given in [75].

Our analysis contains many trade-offs between constants. We do not attempt to make these choices optimally, and no doubt our estimates can be improved upon. Similarly, we do not address issues of precision in computation. While there is no essential difficulty here, to do so introduces a level of tedious detail which we prefer to avoid. We generally assume that computations can be carried out to the required precision, and measure the complexity of the algorithm simply by the number of steps of the random walk.

The plan of Section 4.3 is as follows. In Section 4.3.2 we give the necessary background, and some initial estimates. In Section 4.3.3 we analyse the random walk using a coupling argument. In Section 4.3.4 we describe some improvements, and finally in Section 4.3.5 we discuss the relationship of our work to the literature cited above.

4.3.2 Background and Preliminaries

A number of key parameters arise in the description and analysis of our algorithm, and it is necessary to choose consistent values for these. Rather than attempting to fine-tune everything, we introduce a single constant c which must be chosen large

Table 4.1: The parameters and their relationships.

$$
\begin{array}{rclcrcl}
\lambda & \geq & 1 & \qquad & c & \geq & 2 \\
c\ln n & \geq & 3 & \qquad & c^4 & \leq & R \\
\varrho & = & cnR & \qquad & \sigma & = & r/\sqrt{n} \\
\lambda r & \leq & c^{-4} & \qquad 16\lambda^2 cr\varrho & \leq & 1.
\end{array}
$$

enough for various inequalities to hold. In particular, we require certain events to occur with "very high probability": this is to be interpreted as meaning "with probability at least $1 - n^{-\alpha}$" where the exponent α may be made arbitrarily large by choosing c appropriately.

Three parameters, n, R, and λ, are dictated by the problem instance: n is the dimension, which we assume is at least 2, R is an upper bound on the "sandwiching ratio" of the convex body, and λ is a Lipschitz constant for the derived gauge function (see Section 4.3.2.1 for details). The remaining parameters must be chosen to satisfy certain inequalities, which, for convenience, are gathered together in Table 4.1. To obtain consistent values for these parameters, proceed as follows. First chose c large enough that all the "very high probability" events occur with sufficiently low failure probability; increase R if necessary to achieve $c^4 \leq R$; set

$$
r = \frac{1}{16\lambda^2 c\varrho}
$$

(note that ϱ is determined at this point); finally determine r and σ by reference to the relevant equalities. It can be checked that the remaining inequality is automatically satisfied. In the application to computing volume, $R = \Theta(n)$ and $\lambda = \Theta(1)$, which implies $\varrho = \Theta(n^2)$, $r = \Theta(n^{-2})$, and $\sigma = \Theta(n^{-5/2})$. The parameter r may be interpreted as the typical step-size for the random walk.

4.3.2.1 Convex bodies and gauge functions

Let $K \subseteq \mathbb{R}^n$ be a convex body with $0 \in \mathrm{Int}\, K$ and let ∂K be its boundary. In this area, *convex body* is usually taken to mean a convex set specified by a "well-guaranteed weak separation oracle" [88]. With this technical assumption, it is possible, in polynomial time, to linearly transform K so that it is *well-rounded*, i.e., for some small R, $B \subseteq K \subseteq RB$ where B is the unit ball in $\mathbb{R}^n$. "Small" here means $R = O(n^{3/2})$ [60] if we insist that the inclusion $K \subseteq RB$ be exact, but can be tightened to $R = O(n)$ [88], if we are prepared to allow some negligible proportion of the volume of K to lie outside RB. The linear transformation used to round K alters the volume of the body in a predictable manner.

These situations are often described by the terms *exact* and *approximate sandwiching*. By a recent result of Kannan, Lovász, and Simonovits [74], the bounds on R for exact and approximate sandwiching can be improved to $R = O(n)$ and $R = O(n^{1/2})$, respectively, and these are optimal. However, the latter algorithm requires generating almost uniform points in K, which from our point of view introduces a circularity. We therefore assume only $R = O(n)$. In this section, we take the preprocessing stage as given, since our interest is in analysing a sampling procedure based on a random walk in K. From now on, we can take K to be an arbitrary well-rounded convex set.

We define the *gauge function* of K by $f_K(x) = \inf\{t \geq 0 : x \in tK\}$, and an associated density function[4]

$$F_K(x) = e^{-f_K(x)} \qquad (x \in \mathbb{R}^n).$$

It is well known that f_K is convex and hence F_K is log-concave. A key property of F_K is that sampling from the uniform distribution on K is efficiently reducible to sampling from F_K. One reason for this is that the integral of F_K along a line started at the origin is proportional to the distance from the origin to ∂K along this line—this will be made more precise in the next subsection. Note that $K \subseteq L$ implies $F_K \leq F_L$ pointwise. Also, if we use $\|\cdot\|$ to denote Euclidean distance, $f_B(x) = \|x\|$, so if K is well rounded, we have

$$\frac{\|x\|}{R} \leq f_K(x) \leq \|x\|. \tag{4.10}$$

Note that (4.10) implies finiteness of $\int F_K$.

Since f_K is convex, it possesses at least one *subgradient* $\nabla(x)$ at each point x. That is, for all x, there is a vector $\nabla = \nabla(x)$ such that for all $y \in \mathbb{R}^n$

$$f_K(y) \geq f_K(x) + \nabla \cdot (y - x). \tag{4.11}$$

(The operator "$\cdot$" denotes scalar product.) For $x = 0$, $\nabla = 0$ clearly suffices. For any other point x, we may choose as our ∇ the vector defining a hyperplane $\nabla \cdot z = 1$ through $x/f_K(x) \in \partial K$ and tangential to K, since then $f_K(x) = \nabla \cdot x = 1$ and $f_K(y) \geq \nabla \cdot y$, thus satisfying (4.11). This choice for ∇ satisfies $\|\nabla\| \leq 1$, since $B \subseteq K$. Thus, for any $v \in \partial B$, the (one-sided) *directional derivative* $\delta_v(x)$ of f_K at x in direction v (which is well-defined for any convex function) satisfies

$$|\delta_v(x)| \leq 1. \tag{4.12}$$

4.3.2.2 *Sampling equivalence*

We will show that sampling uniformly from K is equivalent to sampling with density F_K. Specifically, we show that there is a measure-preserving bijection between $\mathbb{R}^n$ and Int K.

Let Γ_n denote the *gamma distribution* on $[0, \infty)$, where Γ_n has probability density $e^{-t}t^{n-1}/\Gamma(n)$. The normalising factor $\Gamma(n)$ here is the *gamma function*, which interpolates $(n-1)!$ at the positive integers. Let $\gamma_n(t) = \int_0^t e^{-u}u^{n-1}\, du/\Gamma(n)$ be the corresponding distribution function. Let $\mathcal{P}_n$ be the distribution on $[0, 1]$ with probability density nt^{n-1} and distribution function t^n.

Let us write $w(x) = \|x\|$ and $b(x) = x/\|x\|$. A point $x \in \mathbb{R}^n \setminus \{0\}$ can be represented uniquely by the polar coordinates (w, b), where $b \in \partial B$ and $w \in (0, \infty)$. Let ξ be the uniform probability measure[5] on ∂B. Then the volume element in $\mathbb{R}^n$ is proportional to $w^{n-1}dw\, d\xi$.

---

[4]We will use the term *density* when the integral is not necessarily 1. If we wish to insist on integral 1, we will write *probability* density.

[5]Measure will always mean probability measure.

Let $h(b) = 1/f_K(b)$ be the distance from the origin to ∂K in the direction b. Note that $f_K(x) = w/h = w(x)/h(b(x))$. Then the uniform probability measure μ_1 on K has element

$$\mathrm{d}\mu_1 = C_1 w^{n-1} \mathrm{d}w\, \mathrm{d}\xi \qquad (b \in \partial B, w \in (0, h]),$$

where clearly $C_1 = 1/\mathrm{Vol}\, K$. Similarly, the measure μ_2 determined by F_K has element

$$\mathrm{d}\mu_2 = C_2 e^{-w/h} w^{n-1} \mathrm{d}w\, \mathrm{d}\xi \qquad (b \in \partial B, w \in (0, \infty)).$$

Integrating with respect to w, the marginal measures ξ_1, ξ_2 on ∂B induced by μ_1 and μ_2 have elements $(C_1/n)h^n\, \mathrm{d}\xi$ and $C_2(n-1)!h^n\, \mathrm{d}\xi$. Comparing, we see that $\xi_1 = \xi_2 = \xi'$ (say), and hence in particular $C_2 = 1/n!\,\mathrm{Vol}\, K$ (cf. [88]).

It follows further that, given a point from μ_2, we can generate a point from μ_1. (The reverse is also true, but we will not elaborate the details.) Note, from the expressions for $\mathrm{d}\mu_1$ and $\mathrm{d}\mu_2$, that the measures ν_1, ν_2 of $f_K(x) = w/h$ are $\mathcal{P}_n$ and Γ_n, respectively. If z is a random point from ν_2, then $\gamma_n(z)^{1/n}$ is a point from ν_1. Thus, if we have a point x_2 from μ_2, we can construct a point x_1 from μ_1 as follows. If $x_2 = 0$ then $x_1 = 0$; otherwise, let $b = x_2/\|x_2\| \in \partial B$, $h = f_K(b)$, $w_2 = \|x_2\|$, and define $x_1 = w_1 b$, where $w_1 = h\gamma_n(w_2/h)^{1/n}$. Then the function χ mapping x_2 to $x_1 = \chi(x_2)$ is a measure-preserving bijection between $\mathbb{R}^n$ and $\mathrm{Int}\, K$.

In practice, we can only sample approximately from F_K. We will take *variation distance* as the measure of distributional approximation. Let $\mu, \hat{\mu}$ be two measures defined on the Borel sets in $\mathbb{R}^n$. (All measures we employ satisfy this assumption, and henceforward we presume the necessary measurability qualification on sets.) Then their variation distance is defined by

$$\mathrm{d}_{\mathrm{TV}}(\mu, \hat{\mu}) = \max_A |\mu(A) - \hat{\mu}(A)|,$$

where $A \subseteq \mathbb{R}^n$ is a . See, for example, Appendix A.1 of [12] for more information. This definition of variation distance is a consistent generalization of that given in Section 1.2.

Variation distance is a metric on measures, but we will extend the notation to random variables. If X, Y are random variables, we will write $\mathrm{d}_{\mathrm{TV}}(X, Y)$ for the variation distance between the measures $\mathbf{P}(X \in \cdot)$, $\mathbf{P}(Y \in \cdot)$ associated with X and Y. Thus, let us denote the approximating measure to μ_2 by $\hat{\mu}_2$, and let $\hat{\mu}_1$ be the resulting approximation to μ_1. Then, since χ is a bijection, we have from the definition of total variation distance that

$$\mathrm{d}_{\mathrm{TV}}(\mu_1, \hat{\mu}_1) = \mathrm{d}_{\mathrm{TV}}(\mu_2, \hat{\mu}_2).$$

Thus the transformation preserves the quality of the approximation.

Note that, in view of this transformation, we can regard a random walk on $\mathbb{R}^n$ converging to the F_K distribution as a random walk on K converging to the uniform distribution. Thus our development below could be described equivalently in terms of K rather than F_K. For example, we define below a distance metric on $\mathbb{R}^n$ to gauge the convergence of the random walk for F_K. The bijection χ then induces a metric on K. We could equally well work with this as a measure for convergence of the random walk in K.

In practice, it is not necessary to determine $w_1 = h\gamma_n(w_2/h)^{1/n}$. We can merely generate an independent uniform $[0, 1]$ variable U and let $w_1 = hU^{1/n}$. The additional generator $\hat{\nu}_1$, approximating ν_1, introduces further error. We have to check that the effect of this on the overall approximation is small. Letting $\tilde{\mu}_1$ be the resulting approximation to μ_1, this follows from:

LEMMA 15.

$$d_{TV}(\mu_1, \tilde{\mu}_1) \leq d_{TV}(\mu_2, \hat{\mu}_2) + d_{TV}(\nu_1, \hat{\nu}_1).$$

Proof. Let $\hat{\xi}'$ be the measure resulting from projecting $\hat{\mu}_2$ onto ∂B. Then it is clear from the definition of variation distance that

$$d_{TV}(\xi', \hat{\xi}') \leq d_{TV}(\mu_2, \hat{\mu}_2).$$

Let $\eta = d_{TV}(\nu_1, \hat{\nu}_1)$. Now, for any $A \subseteq \mathbb{R}^n$ and given $b \in \partial B$, let $A_b = A \cap (0, h)b$. We have $\mu_1(A) = \int_{\partial B} \nu_1(A_b)\,d\xi'$ and

$$\tilde{\mu}_1(A) = \int_{\partial B} \hat{\nu}_1(A_b)\,d\hat{\xi}' \leq \int_{\partial B} (\nu_1(A_b) + \eta)\,d\hat{\xi}'.$$

Hence

$$\tilde{\mu}_1(A) - \mu_1(A) \leq \int_{\partial B} \nu_1(A_b)(d\hat{\xi}' - d\xi') + \eta \leq \int_{B'} (d\hat{\xi}' - d\xi') + \eta,$$

where $B' = \{b \in \partial B : d\hat{\xi}' \geq d\xi'\}$. But

$$\int_{B'} (d\hat{\xi}' - d\xi') = \hat{\xi}'(B') - \xi'(B') \leq d_{TV}(\xi', \hat{\xi}'),$$

from which the lemma follows. $\square$

Thus, if we can generate approximately from F_K, we can sample almost uniformly from K.

4.3.2.3 Modifying F_K

We need to be able to sample from the log-concave distribution defined by F_K. It will be easier to show that we can do this for a slightly modified function, which differs from F_K only at "large" distances from the origin. We will first show that the effect on the sampling distribution will be negligible.

Let $\varrho = cnR$, for some constant $c \geq 2$. For convenience, we will assume without loss of generality that $R \geq c^4$. Then, if X has our target distribution, we have, using (4.10),

$$\mathbf{P}(\|X\| \geq \varrho) = \mathbf{P}(\|X\|/R \geq cn) \leq \mathbf{P}(f_K(X) \geq cn).$$

But we know from Section 4.3.2.2 that $f_K(X)$ has the Γ_n density. Therefore, using Stirling's approximation and Lemma 16 of Section 4.3.2.7 below, we have

$$\mathbf{P}(\|X\| \geq \varrho) \leq 2e^{-cn}(cn)^{n-1}/(n-1)! < (c/e^{c-1})^n. \tag{4.13}$$

Thus, by suitable choice of c, $\|X\| < \varrho$ with very high probability.

In view of this, we may replace $f_K(x)$ outside ϱB by a larger function without significantly altering the distribution determined by f_K, for any polynomial-time sampling scheme. We will choose

$$f_K'(x) = \max\{f_K(x), 2(\|x\| - \varrho)\}.$$

Note that this is convex. If $\|x\| \leq \varrho$, then $f_K'(x) = f_K(x)$, and if $\|x\| \geq 2\varrho$, we have $f_K'(x) = 2(\|x\| - \varrho)$ using (4.10). For our purposes below, the advantage of this function is that it behaves "essentially" like $\|x\|$ at large distances from the origin. Note that the modified probability density F_K' defined by f_K' is very close to the original density F_K: by (4.13), $d_{TV}(F_K, F_K')$ is exponentially small in n. Since the directional derivative of $2(\|x\| - \varrho)$ is 2, we must loosen (4.12) to $|\delta_v(x)| \leq 2$.

Henceforward we will simply write f for f_K' and $F = e^{-f}$, assuming only a condition of the form

$$|\delta_v(x)| \leq \lambda, \tag{4.14}$$

for some given constant $\lambda \geq 1$.

4.3.2.4 Metropolis random walks

Following [73, 88], we construct a discrete time, continuous space Markov random walk (X_t) on $\mathbb{R}^n$ whose stationary distribution is F. Suppose $p(x, x')$ is a (nonnegative) transition density where, for each $x \in \mathbb{R}^n$, we insist that $p(x, x') > 0$ for all x' in some neighbourhood of x and $\int_{\mathbb{R}^n} p(x, x')\, dx' \leq 1$. Then we define

$$\mathbf{P}(X_{t+1} \in x' + dx' \mid X_t = x) \;=\; p(x, x')\, dx', \quad \text{for } x \neq x',$$

and

$$\mathbf{P}(X_{t+1} = x \mid X_t = x) \;=\; 1 - \int_{\mathbb{R}^n} p(x, x')\, dx'.$$

(Here dx' is an infinitesimal neighbourhood of x', and we abuse notation by employing the same symbol for its volume.) Lovász and Simonovits [88] term such a walk "lazy", as there is a positive probability of staying in the same position.

The random walk is *time-reversible* if there exists a probability density π on $\mathbb{R}^n$ such that the *detailed balance* equation

$$\pi(x)p(x, x') = \pi(x')p(x', x), \quad \text{for all } x, x' \in \mathbb{R}^n \text{ with } x \neq x',$$

is satisfied. Then it follows (see [88]) that $\pi(x)$ is the asymptotic density of (X_t). (The density π is certainly invariant; the question is one of uniqueness.)

The random walks in which we are interested arise in the following way. We have a function $q(x, x') : \mathbb{R}^n \times \mathbb{R}^n \to \mathbb{R}$, symmetric in x and x'. We further assume $q(x, \cdot)$ is a probability density for each x, positive in some neighbourhood of x. Suppose $F : \mathbb{R}^n \to (0, \infty)$ is a log-concave function, integrable on $\mathbb{R}^n$, such that the derivative of $f = -\ln F$ is bounded in absolute value by λ in all directions at every point. Let

$$M(x) = e^{-\lambda\|x\|}, \text{ and } A(x, x') = \sqrt{M(x' - x)\frac{F(x')}{F(x)}}.$$

Clearly $0 < A(x, x') \leq 1$ everywhere, since $|f(x') - f(x)| \leq \lambda\|x' - x\|$ is a consequence of the derivative condition. Thus $A(x, x')$ can be interpreted as a probability. Also $A(x, x')/A(x', x) = F(x')/F(x)$, since $M(x' - x) = M(x - x')$. Now define a lazy random walk as above by $p(x, x') = A(x, x')q(x, x')$. Clearly this satisfies

$$F(x)p(x, x') = F(x')p(x', x), \quad \text{for all } x, x' \in \mathbb{R}^n,$$

and hence the asymptotic density of the random walk is $\pi(x) = F(x)/\int F(x)$. This is a modification of the usual Metropolis random walk (see [73]), but is convenient for our proofs. This random walk is easily implemented as follows. If $X_t = x$, we choose a trial point x' with density $q(x, \cdot)$. We accept this point with probability $A(x, x')$. If we accept the point, $X_{t+1} = x'$, otherwise $X_{t+1} = x$. We will let $\mathcal{A}(x, x')$ be the 0–1 indicator variable of the acceptance event, so $\mathbf{P}(\mathcal{A}(x, x') = 1) = A(x, x')$. We will denote the complementary quantities $(1 - A)$, $(1 - \mathcal{A})$ by $\bar{A}$, $\bar{\mathcal{A}}$ respectively.

4.3.2.5 The random walk

Given a log-concave function of the type discussed at the end of Section 4.3.2.3, we define a random walk as in Section 4.3.2.4. We choose $q(x, x')$ to be Gaussian. Let ϕ be the unit Normal density on $\mathbb{R}$, and G_n be the standard Normal density on $\mathbb{R}^n$, i.e.

$$\phi(x) = e^{-x^2/2}/\sqrt{2\pi} \quad \text{and} \quad G_n(x) = \prod_{i=1}^{n} \phi(x_i).$$

Let U be a random variable with density G_n. We take $q(x, x')$ so that $x' = x + \sigma U$, for some parameter σ chosen as in Table 4.1. Then the trial step is to x', and it is accepted with probability $A(x, x')$.

In fact, for convenience, we will modify this random walk slightly. Let $r = \sigma\sqrt{n}$. Then, for large n, r is the approximate step size for X_t. Now, with c as in (4.13), let us redefine the acceptance multiplier $M(x)$ to be

$$M(x) = \exp\left(-\lambda \max\left\{cr, \|x\|\right\}\right).$$

Now $\mathbf{P}(\|x' - x\| > cr) = \mathbf{P}(\|U\|^2 > c^2 n)$. But it is well known that $\frac{1}{2}\|U\|^2$ has the $\Gamma_{n/2}$ density. Hence, using Lemma 16 (in Section 4.3.2.7 below) and Stirling's approximation,

$$\mathbf{P}(\|x' - x\| > cr) \leq \left(c^2/e^{c^2-1}\right)^{n/2} \leq (c/e^{c-1})^n. \tag{4.15}$$

Thus, with a suitable choice of c, we may assume that with very high probability there is no step of size greater than cr during any polynomial number of steps of the random walk. Thus $M(x) = \exp(-\lambda cr)$, where, by suitable choice of σ and hence r, we may assume that $\lambda r \leq c^{-4}$. In Section 4.3.4.2 below, we will describe a further modification to $M(x)$ which asymptotically improves the number of steps required, but for ease of exposition here we use the expression above.

4.3.2.6 Coupling

Previously we have only considered coupling on a finite state space; fortunately, the concepts translate into a far more general setting without any unexpected changes. Details of the extension of the coupling method to more general processes can be found in [57] and [118].

To analyse the convergence rate of our random walk (X_t) to its stationary distribution, we consider a second random walk (Y_t), on the same probability space, with Y_0 taking the stationary distribution. Naturally, to get good bounds, we have to construct the joint process (X_t, Y_t) so that we quickly get $X_t = Y_t$. In order to achieve this here, we use two different couplings depending on $\|X_t - Y_t\|$. The first encourages the two random walks to come closer to each other when they are distant, and the second encourages the random walks to meet if they are sufficiently close.

We wish to have $\mathrm{d_{TV}}(X_t, Y_t) \leq \varepsilon$ for arbitrary ε. However, it is sufficient to have τ', of size polynomial in n, such that $\mathbf{P}(X'_\tau \neq Y'_\tau) \leq {}^1\!/_2$, i.e. $\tau' = \tau({}^1\!/_2)$. Then we may regard t as divided into blocks of length τ', each constituting a coupling "trial" with "success" probability at least ${}^1\!/_2$. Then, after k trials, the probability that coupling has not occurred is at most $({}^1\!/_2)^k$. In more detail, we identify two exceptional events that may occur during the coupling with low probability: the first is that, at some time instant, either random walk (X_t) or (Y_t) takes an unusually large step (this is the event whose complement is denoted $\mathcal{E}$ in Section 4.3.3); the second is that either random walk strays outside the ball $3\varrho B$ of radius 3ϱ. With very high probability, no exceptional event occurs during the entire evolution of the coupling; however, if an exceptional event does occur, we immediately terminate the process and artificially "couple" the random walks (say, by moving both to the origin).

We prove that $\mathbf{P}(X_{(k+1)\tau'} \neq Y_{(k+1)\tau'}) \leq {}^1\!/_2$ (the coupling here may be exceptional or not), provided only that $\|X_{k\tau'}\|, \|Y_{k\tau'}\| \leq 3\varrho$, and other conditioning on $X_{k\tau'}$ and $Y_{k\tau'}$. (Note that we can ensure $\|X_0\| \leq 3\varrho$ by choosing to start the random walk at the origin, and $\|Y_0\| \leq 3\varrho$ with very high probability because Y_0 has the stationary distribution.) Since the evolution of the random walk within each block, conditioned on the entry state of the block, is independent of all previous blocks, the probability of coupling in some block conditioned on coupling not having occurred in previous blocks is at least ${}^1\!/_2$. Hence if $k = \lceil \log_2(1/\varepsilon) \rceil$,

$$\mathrm{d_{TV}}(X_{k\tau'}, Y_{k\tau'}) \leq ({}^1\!/_2)^k + \delta \leq \varepsilon + \delta,$$

where $\delta \leq n^{-\alpha}$ is the probability of an exceptional event, and α can be made arbitrarily small by suitable choice of c (the dependence being $\alpha = \Theta(c^2 \ln n)$). Thus we can guarantee variation distance $\varepsilon = 1/\mathrm{poly}(n)$ by increasing the execution time only by a factor $O(\log n)$. Therefore, we aim below simply to attain a probability ${}^1\!/_2$ of coupling.

4.3.2.7 Technical results

We collect here some simple lemmas which are used in the analysis of Section 4.3.3. All but the most tenacious reader may wish to skip this material, at least on first reading. We first give the proof of a result used in Section 4.3.2.3.

LEMMA 16. *Let Z have the Γ_{k+1} density, and let $\mu \geq 2k$. Then*

$$e^{-\mu}\mu^k/k! \leq \mathbf{P}(Z \geq \mu) \leq 2e^{-\mu}\mu^k/k!$$

Proof. The left-hand inequality is easily proved by integration by parts. Also, since $\mu \geq 2k$,

$$
\begin{aligned}
\mathbf{P}(Z \geq \mu) &= \frac{1}{k!}\int_\mu^\infty e^{-t}t^k\,\mathrm{d}t \\
&\leq \frac{1}{k!}e^{-\mu/2}\mu^k\int_\mu^\infty e^{-t/2}\,\mathrm{d}t \\
&= 2e^{-\mu}\mu^k/k!
\end{aligned}
$$

$\square$

We now state a very simple bound on $\ln n$, which we use repeatedly without comment. It may be proved by elementary calculus.

LEMMA 17. *If $\alpha > 0$, then $\ln n/n^\alpha \leq 1/(e\alpha)$.*

We next prove two simple approximation results.

LEMMA 18. *Let $a \geq \frac{1}{4}$, and $b \geq 0$. Then $\sqrt{a+b} \leq \sqrt{a} + b$.*

Proof. This follows from $a + b \leq a + 2\sqrt{a}\,b \leq (\sqrt{a} + b)^2$, on taking square roots. We require only $2\sqrt{a} \geq 1$, i.e. $a \geq \frac{1}{4}$. $\square$

COROLLARY 11. *Let $a \geq \frac{1}{4}$, and $b \geq 0$. Then $\sqrt[4]{a+b} \leq \sqrt[4]{a} + b$.*

Proof. Applying Lemma 18 twice

$$\sqrt[4]{a+b} \leq \sqrt{\sqrt{a}+b} \leq \sqrt[4]{a} + b,$$

where the first application is valid for $a \geq {}^1\!/_4$, and the second for $a \geq {}^1\!/_{16}$. $\square$

LEMMA 19. *If $|z| \leq {}^1\!/_2$, then $\sqrt{1-z} \leq 1 - {}^1\!/_2 z - \frac{1}{10}z^2$.*

Proof. Since $1 - {}^1\!/_2 z - \frac{1}{10}z^2$ is positive for $|z| \leq 1$, we may obtain an equivalent inequality by squaring both sides. Simplifying this gives $z^2 + 10z + 5 \geq 0$. This is satisfied for all $|z| \leq {}^1\!/_2$. $\square$

We now prove a simple bound on the tail of the Normal distribution.

LEMMA 20. *Let $v \sim N(0,\sigma^2)$ and $\ell = c\sigma \ln n$, then $\mathbf{P}(|v| > \ell) \leq n^{-c^2 \ln n/2}$.*

Proof.

$$
\begin{aligned}
\mathbf{P}(|v| > \ell) &= \frac{2}{\sigma}\int_\ell^\infty \phi(t/\sigma)\,\mathrm{d}t \leq \frac{2}{\sigma\ell}\int_\ell^\infty t\phi(t/\sigma)\,\mathrm{d}t \\
&= \frac{2\sigma}{\ell}\phi(\ell/\sigma) \leq n^{-c^2 \ln n/2},
\end{aligned}
$$

provided $c\ln n \geq 2$. $\square$

Next we prove two easy results to deal with mild conditioning in the analysis.

LEMMA 21. *Let Z be a real valued random variable, symmetrically distributed about the origin, and f be a function satisfying $\mathbf{E}\left(|f(Z)|\right) < \infty$. For $t \geq 0$, let E_t be the event $|Z| \leq t$. Then if f is an*

(i) odd function, $\mathbf{E}\left(f(Z) \mid E_t\right) = 0$,

(ii) even convex function, $\mathbf{E}\left(f(Z) \mid E_t\right) \leq \mathbf{E}\left(f(Z)\right)$.

Proof. The first statement follows by symmetry, the second by noting that $f(z_1) \leq f(z_2)$ for $|z_1| \leq |z_2|$. $\square$

LEMMA 22. *Let $Z \sim N(0, \sigma^2)$, and for $t \geq 3\sigma$ let E_t be the event $|Z| \leq t$. Then $\mathbf{E}\left(Z^2 \mid E_t\right) \geq 0.97\sigma^2$.*

Proof. It is sufficient to prove the lemma for $\sigma = 1$. Then $\mathbf{E}\left(Z^2 \mid E_t\right) = 1 - t\phi(t)/(\Phi(t) - 1/2)$, where Φ denotes the standard Normal distribution function. This expression is increasing for all $t > 0$, and putting $t = 3$ gives $(1 - 0.0133/0.4987) > 0.97$. $\square$

Finally we give a bound on a simple function.

LEMMA 23. *For $z \leq \frac{1}{10}$, ze^{-z} is increasing and bounded above by $z - \frac{19}{20}z^2$.*

Proof. The first claim may be proved by elementary calculus. If $z \leq 0$, the second follows from $e^{|z|} \geq 1 + |z|$. If $z > 0$, note that the series for e^{-z} is alternating with deceasing terms for $z < 1$, and hence

$$e^{-z} < 1 - z + 1/2z^2 = 1 - z(1 - 1/2z) \leq 1 - \frac{19}{20}z,$$

from which the result follows. $\square$

4.3.3 Analysis of the Random Walk

We assume $n \geq 2$. We present the coupling analysis in three sections. In Section 4.3.3.1 we prove boundedness, in Section 4.3.3.2 we show approximate coupling, and in Section 4.3.3.3 we show that exact coupling will eventually occur.

4.3.3.1 *Boundedness of the walk*

It will be necessary for us to be sure that with very high probability the random walks do not stray too far from the origin. For Y_t (started in the stationary distribution) this can be established easily by a calculation like that leading to (4.13). However for X_t (started at an arbitrary point) it is not so straightforward. To bound this we consider the induced stochastic process $\Xi_t = \|X_t\|$. We show, assuming only $\Xi_0 < 2\varrho$, that $\Xi_t < 3\varrho$ with very high probability for all t from 0 up to some (arbitrary) polynomial bound.

Let us call any set of consecutive values of t for which $\Xi_t > 2\varrho$ an *excursion*. We wish to show that any excursion does not include values of $\Xi_t > 3\varrho$. Suppose $X_t = x$

with $\Xi_t = \xi \geq 2\varrho$, and let $s = \Xi_{t+1} - \Xi_t$. Then we know that $f(x) = 2(\xi - \varrho)$, and hence we will assume $\lambda \geq 2$ for the remainder of this section. (The argument may be modified, with a weaker conclusion, so as to hold for any positive λ.) Also, if $x' = x + \sigma U$ is the trial point as in Section 4.3.2.5, then letting $\xi' = \|x'\|$ and $s' = \xi' - \xi$, we have $F(x')/F(x) = e^{-2s'}$, if $\xi' \geq 2\varrho$.

Now write $u = x/\xi$, and let $v = \sigma U \cdot u$ and $w = \sqrt{\sigma^2 \|U\|^2 - v^2}$. Then

$$\xi' = \sqrt{(\xi + v)^2 + w^2} = \xi\sqrt{(1 + v/\xi)^2 + (w/\xi)^2}.$$

Let $\ell = c\sigma \ln n$. Then $\ell = (c/\lambda)\lambda r(\ln n/\sqrt{n}) \leq (\ln 7)/16\sqrt{7} \leq \frac{1}{21}$. Denote the events $|v| \leq \ell$ and $w < 2cr/3$ by $\mathcal{E}_1$ and $\mathcal{E}_2$, respectively, and let $\mathcal{E} = \mathcal{E}_1 \cap \mathcal{E}_2$. With very high probability, by Lemma 20 and equation (4.15), $\mathcal{E}$ occurs at every step for polynomially many steps, provided c is large enough. It is worth noting that $\mathcal{E}$ implies $\sigma\|U\| \leq cr\sqrt{4/9 + (\ln n)^2/n} < cr$. Thus, if $\mathcal{E}$ occurs, the acceptance indicator $\mathcal{A}$ has

$$\mathbf{P}(\mathcal{A} = 1 \mid \mathcal{E}) = A = \sqrt{e^{-\lambda cr}e^{-2s'}} = e^{-(c'+s')},$$

where $c' = \lambda cr/2 \leq \frac{1}{16}$.

Assuming the same event, $v/\xi \geq -\ell/2cnR = -r\ln n/2Rn^{3/2} \geq -1/4000$. Hence we may apply Lemma 18 to give

$$\xi' \leq \xi(1 + v/\xi + (w/\xi)^2) = \xi + v + w^2/\xi \leq \xi + v + \varepsilon,$$

for some constant ε, where $0 \leq \varepsilon < 4(cr)^2/18cnR \leq \sigma^2/36$. Also it follows that $\xi' \geq \xi + v$, and hence

$$v \leq s' \leq v + \varepsilon.$$

Since $\varepsilon < \ell/1000$, we have $|s| \leq |s'| \leq \ell + \varepsilon < 1.001\ell \leq \frac{1}{20}$. Now, using Lemmas 21, 22, and 23,

$$
\begin{aligned}
\mathbf{E}\left(s \mid \mathcal{E}\right) &= \mathbf{E}\left(s'e^{-(c'+s')} \mid \mathcal{E}\right) \\
&\leq e^{-c'}\mathbf{E}\left((v+\varepsilon)e^{-(v+\varepsilon)} \mid \mathcal{E}_1\right) \\
&\leq e^{-c'}\mathbf{E}\left((v+\varepsilon) - 0.95(v+\varepsilon)^2 \mid \mathcal{E}_1\right) \\
&= e^{-c'}\mathbf{E}\left(\varepsilon(1 - 0.95\varepsilon) + (1 - 1.9\varepsilon)v - 0.95v^2 \mid \mathcal{E}_1\right) \\
&\leq e^{-c'}(\varepsilon - 0.92\sigma^2) \\
&\leq -0.83\sigma^2,
\end{aligned}
$$

using $\varepsilon \leq \sigma^2/36$ and $c' \leq 1/16$. Hence, conditionally on $\mathcal{E}$, $|\Xi_{t+1} - \Xi_t| < 1.001 c\sigma \ln n$ and $\mathbf{E}(\Xi_{t+1} - \Xi_t \mid \Xi_t) \leq -0.83\sigma^2$. Suppose the excursion starts at $t = a + 1$, so $\Xi_a \leq 2\varrho$, and $k_0 = \lfloor c^4(\ln n)^4/2\sigma^2 \rfloor$. Then, since the process Ξ_t is a supermartingale, Hoeffding's inequality for martingales [63] (see also [93]) gives

$$\mathbf{P}\left(\exists k, 1 \leq k \leq k_0 : \Xi_{a+k} - \Xi_a + 0.83\sigma^2 k > 0.4c^4(\ln n)^4\right) \leq n^{-(c^2 \ln n)/7}.$$

$$(4.16)$$

However, (4.16) implies

$$\mathbf{P}(\Xi_{a+k_0} - \Xi_a > 0) \leq n^{-(c^2 \ln n)/7}.$$

Thus, with very high probability, the excursion must have already ended at step $(a + k_0)$. For $k < k_0$, (4.16) implies

$$\mathbf{P}\left(\exists k < k_0 : \Xi_{a+k} > 2\varrho + 0.4c^4(\ln n)^4\right) \leq n^{-(c^2 \ln n)/7}.$$

But $0.4c^4(\ln n)^4 < 2c^4 n \leq 2nR \leq \varrho$. Thus each excursion lies entirely in the ball $3\varrho B$, except with probability $n^{-(c^2 \ln n)/7}$. Since the algorithm can only perform a polynomial number of excursions, we can choose c so that $3\varrho B$ includes the whole walk with very high probability.

4.3.3.2 Bringing the random walks close

We have two random walks (X_t), (Y_t) as in Section 4.3.2.5, with Y_0 chosen randomly from the equilibrium density π. We will couple them in the following way. Suppose σU is the trial step. Let $u_1, u_2, \ldots, u_n$ be an orthonormal basis for $\mathbb{R}^n$ such that $u_1 = (Y_t - X_t)/\|Y_t - X_t\|$. Then $U = \sum_{i=1}^n U_i u_i$, where U_i has the density $G_1 = \phi$. Let us write $U' = \sum_{i=2}^n U_i u_i$, so U' has the density G_{n-1}. It is a fundamental property of the multivariate Normal distribution that U' and U_1 are independent. Then the trial points are generated by

$$X_t' = X_t + \sigma U_1 u_1 + \sigma U', \qquad Y_t' = Y_t - \sigma U_1 u_1 + \sigma U'.$$

We accept the trial points *independently* with probabilities $A(X_t, X_t')$ and $A(Y_t, Y_t')$, respectively. If we accept X_t', then $X_{t+1} = X_t'$, otherwise $X_{t+1} = X_t$ and similarly for Y_t'. Thus the two random walks are coupled so that, if the two steps are accepted, their steps will be reflections of each other in the hyperplane of points equidistant from X_t and Y_t. This has similarities to the coupling given in [85].

We will show that, with this coupling, X_t and Y_t have a tendency to converge. Specifically, we show that, for some metric d on $\mathbb{R}^n$, $\mathbf{E}\left(d(X_t, Y_t)\right)$ decreases. The metric we choose is the *square-root* of Euclidean distance, $d(X_t, Y_t) = \sqrt{\|X_t - Y_t\|}$. The rationale for this metric is to downweight larger values of the "neutral" Euclidean distance. We will write $d_t = d(X_t, Y_t)$, and let D denote $d_t^2 = \|Y_t - X_t\|$. The bottom line will be that in some (fixed) polynomial number of steps (in fact $O(n^9 \log n)$ steps) the walks will come close with probability at least $1/2$. From there, a second coupling, described in Section 4.3.3.3, will take over to complete the job.

In notation similar to that in Section 4.3.3.1, let us write $v = \sigma U_1$ and $w = \|\sigma U'\|$, and let us denote the event $|v| \leq c\sigma \ln n$ by $\mathcal{E}_1$, the event $w < 2cr/3$ by $\mathcal{E}_2$, and $\mathcal{E}_1 \cap \mathcal{E}_2$ by $\mathcal{E}$. Also, abbreviating the indicator variables for acceptance to $\mathcal{A}_X$, $\mathcal{A}_Y$, let

$$\begin{aligned} I &= \mathcal{A}_X + \mathcal{A}_Y = 1 + \mathcal{A}_X \mathcal{A}_Y - \bar{\mathcal{A}}_X \bar{\mathcal{A}}_Y, \\ J &= |\mathcal{A}_X - \mathcal{A}_Y| = |\bar{\mathcal{A}}_X - \bar{\mathcal{A}}_Y|, \end{aligned}$$

where $I \in \{0, 1, 2\}$ and $J \in \{0, 1\}$. Then it follows that

$$d_{t+1} = \sqrt[4]{(d_t^2 - vI)^2 + w^2 J}.$$

We will bound $\mathbf{E}\left(d_{t+1} \mid X_t, Y_t, \mathcal{E}\right)$ as a function of $d_t = d(X_t, Y_t)$. From now on, since all probabilities and expectations will be conditioned on X_t and Y_t, we drop explicit references to those quantities.

Note that, with very high probability, $\mathcal{E}$ occurs at every step for sufficiently many steps, and no step size is larger than cr. Now $\mathcal{E}$ implies

$$D - vI \geq D - 2c\sigma \ln n \geq \frac{1}{2}D,$$

provided we do not have the event $d_t < 2\sqrt{c\sigma \ln n}$. Let us call this event $\mathcal{C}$. We will say the walks are "close" when $\mathcal{C}$ occurs. Now if $\mathcal{C}$ does not occur, using Corollary 11 and Lemma 19,

$$
\begin{aligned}
d_{t+1}/d_t &= \sqrt[4]{(1 - vI/D)^2 + w^2 J/D^2} \\
&\leq \sqrt{1 - vI/D + w^2 J/D^2} \\
&\leq 1 - \frac{1}{2}vI/D - v^2 I^2/10D^2 + w^2 J/D^2. \qquad (4.17)
\end{aligned}
$$

We have to bound the terms on the right side of (4.17) conditioned by $\mathcal{E}$ and $\bar{\mathcal{C}}$.

Note that if $\mathcal{E}$ occurs then the step size is less than cr, and hence, for $A = A_X, A_Y$,

$$e^{-\lambda cr} \leq A \leq 1, \qquad 0 \leq \bar{A} \leq 1 - e^{-\lambda cr} \leq \lambda cr.$$

Thus, using Lemmas 21(ii) and 22,

$$
\begin{aligned}
\mathbf{E}\left(w^2 J \mid \mathcal{E}, \bar{\mathcal{C}}\right) &\leq \mathbf{E}\left(w^2(\bar{A}_X + \bar{A}_Y) \mid \mathcal{E}\right) \\
&\leq 2\lambda cr \mathbf{E}\left(w^2 \mid \mathcal{E}_2\right) \leq 2\lambda cr \mathbf{E}\left(w^2\right) < 2\lambda cr^3, \qquad (4.18)
\end{aligned}
$$

and, provided $c \ln n \geq 3$,

$$
\begin{aligned}
\mathbf{E}\left(v^2 I^2 \mid \mathcal{E}, \bar{\mathcal{C}}\right) &= \mathbf{E}\left(v^2(A_X + A_Y + 2A_X A_Y) \mid \mathcal{E}\right) \\
&\geq 2e^{-c\lambda r}(1 + e^{-c\lambda r})\mathbf{E}\left(v^2 \mid \mathcal{E}_1\right) \geq 3r^2/n, \qquad (4.19)
\end{aligned}
$$

since $c\lambda r \leq \frac{1}{8}$ and $\mathbf{E}\left(v^2 \mid \mathcal{E}_1\right) \geq 0.97\sigma^2$.

It remains to consider $\mathbf{E}\left(vI \mid \mathcal{E}, \bar{\mathcal{C}}\right)$. Now, since $\mathbf{E}\left(v \mid \mathcal{E}\right) = \mathbf{E}\left(v \mid \mathcal{E}_1\right)$, we have using Lemma 21 that

$$
\begin{aligned}
\mathbf{E}\left(vI \mid \mathcal{E}, \bar{\mathcal{C}}\right) &= \mathbf{E}\left(v + vA_X A_Y - v\bar{A}_X \bar{A}_Y \mid \mathcal{E}\right) \\
&= \mathbf{E}\left(vA_X A_Y \mid \mathcal{E}\right) - \mathbf{E}\left(v\bar{A}_X \bar{A}_Y \mid \mathcal{E}\right).
\end{aligned}
$$

We will bound these two contributions separately. We do this by arguing conditionally on both $\mathcal{E}_1$ and the value of w (given that this satisfies $\mathcal{E}_2$). Since the bounds we obtain are independent of w, we may infer the same bound for conditioning on $\mathcal{E}$. In this respect, for A being either A_X or A_Y, the notation $A(v)$ will mean "A considered as a function of v for a given fixed w".

Thus let $p(v)$ be the conditional density of v given $\mathcal{E}$, and first consider the second term. Then

$$
\begin{aligned}
\mathbf{E}\left(v\bar{A}_X\bar{A}_Y \mid \mathcal{E}_1, w\right) &= \int_{-\infty}^{\infty} v\bar{A}_X(v)\bar{A}_Y(v)p(v)dv \\
&= \int_0^{\infty} v(\bar{A}_X(v)\bar{A}_Y(v) - \bar{A}_X(-v)\bar{A}_Y(-v))p(v)dv \\
&= 2\int_0^{\infty} \frac{\bar{A}_X(v)\bar{A}_Y(v) - \bar{A}_X(-v)\bar{A}_Y(-v)}{2v} v^2 p(v)dv \\
&= 2\int_0^{\infty} \left[\frac{\mathrm{d}(\bar{A}_X(v)\bar{A}_Y(v))}{dv}\right]_\theta v^2 p(v)dv,
\end{aligned}
$$

where $-v \le \theta \le v$. Note that if $\bar{A}_X(v)\bar{A}_Y(v)$ is not differentiable everywhere we may consider a suitably close approximant (see e.g. [54]). Now

$$
\frac{\mathrm{d}(\bar{A}_X(v)\bar{A}_Y(v))}{dv} = -\left(\bar{A}_X(v)A_Y(v)\frac{\mathrm{d}(\ln A_Y(v))}{dv} + \bar{A}_Y(v)A_X(v)\frac{\mathrm{d}(\ln A_X(v))}{dv}\right)
$$

and hence, using the bound λ on the directional derivative of f,

$$
\left|\frac{\mathrm{d}(\bar{A}_X(v)\bar{A}_Y(v))}{dv}\right| \le \lambda^2 cr.
$$

Thus, using Lemma 21(ii),

$$
\begin{aligned}
\mathbf{E}\left(v\bar{A}_X\bar{A}_Y \mid \mathcal{E}_1, w\right) &\le 2\lambda^2 cr \int_0^{\infty} v^2 p(v)dv \\
&= \lambda^2 cr \mathbf{E}\left(v^2 \mid \mathcal{E}_1\right) \\
&\le \lambda^2 cr\sigma^2 = \lambda^2 cr^3/n. \qquad (4.20)
\end{aligned}
$$

Now let $g(v) = \sqrt{e^{-\lambda cr}F(X_t + \sigma U' + vu_1)}$. Then

$$
g(D - v) = \sqrt{e^{-\lambda cr}F(X_t + \sigma U' + Du_1 - vu_1)} = \sqrt{e^{-\lambda cr}F(Y_t + \sigma U' - vu_1)}.
$$

Note that $g(v)$ is a log-concave function (of one argument), and

$$
A_X(v) = \frac{g(v)}{\sqrt{F(X_t)}}, \qquad A_Y(v) = \frac{g(D - v)}{\sqrt{F(Y_t)}}.
$$

Therefore, writing $C = (F(X_t)F(Y_t))^{-1/2}$, we have

$$
\begin{aligned}
\mathbf{E}\left(vA_X A_Y \mid \mathcal{E}_1, w\right) &= \int_{-\infty}^{\infty} vA_X(v)A_Y(v)p(v)\,dv \\
&= C\int_{-\infty}^{\infty} vg(v)g(D - v)p(v)\,dv \\
&= C\int_0^{\infty} (g(v)g(D - v) - g(-v)g(D + v))v\,p(v)\,dv, \\
&\ge 0, \qquad (4.21)
\end{aligned}
$$

since the integrand is non-negative by the log-concavity of $g(v)$. To show this, let $t(v) = -\ln g(v)$, so $t(v)$ is convex. We have now to show that $t(v) - t(-v) \leq t(D+v) - t(D-v)$. But this is clear since

$$t(v) - t(-v) = \int_{-v}^{v} t'(\eta)\,d\eta, \qquad t(D+v) - t(D-v) = \int_{D-v}^{D+v} t'(\eta)\,d\eta,$$

for some non-decreasing function $t'(\eta)$.

We note in passing that it is only to establish (4.21) that we require log-concavity of F. All our other estimates simply require suitable "smoothness" of F. On the other hand, for something like (4.21) to be true, it appears we need a property close to log-concavity. It is also the proof of (4.21) which necessitates our departure from the "usual" Metropolis process, since we have to factor out the different contributions from the points X_t, Y_t.

From inequalities (4.20) and (4.21), the definition of I, and the independence of the two acceptance probabilities, we now have

$$\mathbf{E}\left(vI \mid \mathcal{E}\right) \geq -\lambda^2 cr^3/n. \tag{4.22}$$

Therefore, putting (4.18), (4.19), and (4.22) into (4.17), we have

$$\begin{aligned}
\mathbf{E}\left(d_{t+1}/d_t \mid \mathcal{E}, \bar{\mathcal{C}}\right) \;&\leq\; 1 + \frac{\lambda^2 cr^3}{2nD} - \frac{3r^2}{10nD^2} + \frac{2\lambda cr^3}{D^2} \\
&=\; 1 - \frac{r^2}{10nD^2}\left(3 - 5\lambda^2 crD - 20\lambda crn\right) \\
&\leq\; 1 - \frac{r^2}{10nD^2},
\end{aligned}$$

provided $5\lambda^2 crD + 20\lambda crn \leq 2$. This will be true if $16\lambda^2 cr\varrho \leq 1$, using $R \geq c^4$ and the fact that, with very high probability, $D \leq 3\varrho + 3\varrho = 6\varrho$ from Section 4.3.3.1. Thus let us take

$$r = \frac{1}{16c\lambda^2\varrho} = \frac{1}{16c^2\lambda^2 nR}. \tag{4.23}$$

Thus if $R = O(n)$, $r = \Omega(1/n^2)$. Now, using $D \leq 6\varrho$,

$$\begin{aligned}
\mathbf{E}\left(d_{t+1}\right) \;&\leq\; \mathbf{E}\left(d_t\right)\left(1 - \frac{1}{95000nc^2\lambda^4\varrho^4}\right) \\
&=\; \mathbf{E}\left(d_t\right)\left(1 - \frac{1}{95000n^5c^6\lambda^4 R^4}\right), \tag{4.24}
\end{aligned}$$

conditioned on $\mathcal{E}$ and $\bar{\mathcal{C}}$ throughout. The reduction factor is $(1 - \Omega(1/n^9))$, assuming $R = O(n)$. Therefore, since $d_0 = O(n)$ and $\sigma = \Omega(n^{-5/2})$, in $O(n^9 \log n)$ steps we must either have encountered $\mathcal{C}$ or $\mathbf{E}\left(d_t\right) \leq \frac{1}{10}\sqrt{c\sigma \ln n}$. In the latter case, by Markov's inequality, $\mathbf{P}(\bar{\mathcal{C}}) \leq \frac{1}{20}$. Hence, in a sequence of $O(n^9 \log n)$ steps, there is probability at least $\frac{19}{20}$ that $\mathcal{C}$ holds at some step.

4.3.3.3   Making the random walks meet

When $\mathcal{C}$ occurs, i.e. $d_t < 2\sqrt{c\sigma \ln n}$, we switch to a different coupling. Note that X_t and Y_t may still be relatively far apart compared to the expected movement in any direction in a single step. Therefore we choose instead to correlate a *sequence* of steps in order to have sufficient probability that $X_t = Y_t$ at the end of the sequence.

Thus suppose $\mathcal{C}$ occurs at step τ', but $\bar{\mathcal{C}}$ held at $(\tau' - 1)$. Let $s_{X,t} = X_t' - X_t$, $s_{Y,t} = Y_t' - Y_t$ denote the trial steps in the two walks. Let $k = \lceil (4c \ln n)^2 \rceil$, and

$$S_X = \sum_{t=\tau'}^{\tau'+k-1} s_{X,t}, \qquad Q_X = X_\tau' + S_X, \qquad (4.25)$$

and similarly S_Y, Q_Y. Note that $X_{\tau'+k} = Q_X$ and $Y_{\tau'+k} = Q_Y$ if all trial steps are accepted for $t = \tau', \tau' + 1, \ldots, \tau' + k - 1$.

Now S_X, S_Y each has density $G_n(x/\sigma\sqrt{k})$. We will couple S_X, S_Y. Then $s_{X,t}, s_{Y,t}$ are generated independently with density $\phi(x/\sigma)$, subject to the sum condition (4.25).

As in Section 4.3.3.2, let $u_1, u_2, \ldots, u_n$ be an orthonormal basis such that $u_1 = (Y_\tau' - X_\tau')/D$, where $D = \|Y_\tau' - X_\tau'\|$. Suppose $S_X = (v, v_2, \ldots, v_n)$ in this co-ordinate system, where $v, v_2, \ldots, v_n$ are independent with probability density $\psi(x)$ proportional to $\phi(x/\sigma\sqrt{k})$. We now take $S_Y = (-v', v_2, \ldots, v_n)$, where v' will be defined in terms of v. Note that we have $Q_X = Q_Y$ if and only if $v' = D - v$. Thus, for given v, define v' by

$$v' = \begin{cases} D - v, & \text{with probability } \min\{1, \psi(D - v)/\psi(v)\}, \\ v, & \text{otherwise.} \end{cases}$$

Intuitively, we are reflecting the tail of the density $\psi(v)$ about the point $v = \frac{1}{2}D$. Then v' has probability density function

$$\psi(D - v')\min\{1, \psi(v')/\psi(D - v')\}$$
$$+ \psi(v')(1 - \min\{1, \psi(D - v')/\psi(v')\}) = \psi(v').$$

Hence, since $\psi(v') = \psi(-v')$, $-v'$ has density ψ, as required. Also

$$
\begin{aligned}
\mathbf{P}(Q_X = Q_Y) = \mathbf{P}(v' = D - v) \; &= \; \int_{-\infty}^{\infty} \min\{\psi(D - v), \psi(v)\}dv \\
&= \; 2\int_{-\infty}^{-D/2} \psi(v)dv \\
&= \; 2\Phi\left(\frac{-D}{2\sigma\sqrt{k}}\right) \\
&\geq \; 2\Phi(-\tfrac{1}{2}) > 0.616.
\end{aligned}
$$

We must now consider the probability that all steps accept. To maximize this, we choose to make the random variables $\mathcal{A}_X, \mathcal{A}_Y$ dependent at each step. Their values are generated as follows. Let w be a random variable, independent for each step, with distribution uniform on $[0, 1]$. Then $\mathcal{A}_X = 1$ if $w \le A_X$, otherwise $\mathcal{A}_X = 0$. The value of $\mathcal{A}_Y$ is chosen similarly, using the same w. The advantage of this is that now

$$\mathbf{P}(\mathcal{A}_X = \mathcal{A}_Y = 1 \mid \mathcal{E}) = \min\{A_X,\ A_Y\} \ge e^{-c\lambda r}.$$

Now if $\mathcal{G}$ is the event that all trial steps are accepted in both walks for $t = \tau', \tau' + 1, \ldots, \tau' + k - 1$, we have

$$\mathbf{P}(\mathcal{G}) \ge \left(e^{-c\lambda r}\right)^k \ge e^{-1/14} > 0.931,$$

using the value of r given in (4.23). Hence we have

$$\mathbf{P}(X_{\tau'+k} = Y_{\tau'+k}) = \mathbf{P}(Q_X = Q_Y)\mathbf{P}(\mathcal{G}) > 0.616 \times 0.931 > 0.573.$$

The number of steps, $O((\log n)^2)$, is asymptotically negligible compared with the number required in Section 4.3.3.2 to achieve $\mathbf{P}(\mathcal{C}) > \frac{19}{20}$. Thus, after $O(n^5 R^4 \log n)$ steps, the probability that the walks have failed to couple is at most $\frac{1}{20} + \frac{19}{20}(1 - 0.573) < 0.46$.

This completes the analysis of the algorithm. Putting $R = O(n)$, our "mixing time" is $O^*(n^9)$, where $O^*(\cdot)$ is the notation which hides factors of $\log n$. This is larger than is known to be possible [75]. Note that the mixing time is $O^*(n\varrho^2/r^2) = O^*((\varrho/\sigma)^2)$, By comparison with the one-dimensional random walk, this is of the optimal form. The problem is that the step size, r, is too small. We attempt to offset this difficulty in the next section.

4.3.4 Improvements

In this section we describe two refinements which reduce the required number of steps of the random walk from $O^*(n^9)$ to $O^*(n^{6.5})$.

4.3.4.1 *A faster simulation of the random walk*
An idea similar to that used to couple the walks in Section 4.3.3.3 can be employed to simulate the random walk with fewer actual steps. A difficulty with our walk is that $r = \Theta(1/\varrho)$ is smaller than values known to be achievable [88]. However, the fact that our trial steps are so small means that almost all will be accepted, and hence we may simulate many steps simultaneously in a larger "step".

Suppose at step t, we attempt to perform k steps, where k is a power of 2. We will call this a *k-step*. Denote the current k-step by S. Let S be the *vector sum* of the k trial steps in S, as in Section 4.3.3.3. We generate the components of S as independent $N(0, k\sigma^2)$. If a is the probability that all these trial steps are accepted, then we have

$$a = \sqrt{e^{-k\lambda cr} F(X_t + S)/F(X_t)},$$

since all other factors cancel. Thus we accept S with probability a. If we accept, then we update t to $t + k$ and consider the next k-step.

If however, with probability $(1 - a) \le k\lambda cr$, we do not accept S, we know that at least one constituent step was rejected. We will say S *rejects*. We subdivide S

into two $^{1}/_{2}k$-steps $\mathcal{S}_1, \mathcal{S}_2$ (in that order). We generate the corresponding vector sums S_1, S_2 subject to the condition $S_1 + S_2 = S$. Suppose s_1 is the ith component of S_1, and s the ith component of S. Then s_1 has conditional density

$$\frac{(e^{-s_1^2/k\sigma^2}/\sqrt{\pi k\sigma^2})(e^{-(s-s_1)^2/k\sigma^2}/\sqrt{\pi k\sigma^2})}{e^{-s^2/2k\sigma^2}/\sqrt{2\pi k\sigma^2}} = \frac{1}{\sqrt{^{1}/_{2}\pi k\sigma^2}}e^{-2(s_1-s/2)^2/k\sigma^2}.$$

Thus $s_1 \sim N(^{1}/_{2}s, ^{1}/_{4}k\sigma^2)$. Hence we can generate S_1, S_2. Now let

$$a_1 = \sqrt{e^{-k\lambda cr/2}F(X_t + S_1)/F(X_t)}$$

be the unconditioned acceptance probability for $\mathcal{S}_1$. We must condition on the event that $\mathcal{S}$ rejects. Let q be the conditional probability that $\mathcal{S}_1$ does not reject. Then clearly $q = (a_1 - a)/(1 - a)$. We use q to decide (by random generation) if $\mathcal{S}_1$ rejects. If not, we can set $X_{t+k/2} = X_t + S_1$ and consider $\mathcal{S}_2$ (with vector sum S_2), conditional that it rejects. The base case of the recursion, $k = 1$, is clearly a (single) step.

If $\mathcal{S}_1$ rejects, then we (recursively) halve it conditional on this event. Eventually, using this recursion, we compute $X_{t+k/2}$. Then we consider $\mathcal{S}_2$, with vector sum S_2, now as an unconditioned $^{1}/_{2}k$-step. Hence we ultimately compute X_{t+k}.

To analyse this procedure, let $t_u(k)$ denote the supremum, over all possible states of X_t, of the expected total number of k'-steps which are accepted within $\mathcal{S}$, for all $k' \leq k$. This effectively determines an upper bound for the execution time for $\mathcal{S}$. Let $t_c(k)$ be this number conditional that $\mathcal{S}$ rejects. For notational convenience, let $\beta = \lambda cr$. Then, from the above,

$$\begin{aligned}
t_u(k) &\leq a + (1-a)t_c(k) \leq 1 + k\beta t_c(k) \\
t_c(k) &\leq q(1 + t_c(^{1}/_{2}k)) + (1-q)(t_c(^{1}/_{2}k) + t_u(^{1}/_{2}k)) \\
&\leq 1 + t_c(^{1}/_{2}k) + t_u(^{1}/_{2}k).
\end{aligned}$$

Hence we have

$$t_c(k) \leq 2 + (1 + k\beta)t_c(^{1}/_{2}k) \leq 2 + e^{k\beta}t_c(^{1}/_{2}k). \tag{4.26}$$

Now (4.26) implies $t_c(k) \leq 2e^{2k\beta}\log_2 k$, as is easily shown by induction. (Note $t_c(1) = 0$.) Hence

$$t_u(k) \leq 1 + 2e^{2k\beta}k\beta\log_2 k.$$

Let $k = 2^{\lfloor \log_2(1/\beta)\rfloor}$. Note that $k \leq 1/\beta$ and $k = \Omega(1/r)$. Then

$$t_u(k) \leq 1 + 2e^2\log_2(1/\beta) = O(\log(1/r)).$$

Thus we trade a factor $1/r$ in the number of steps for $\log(1/r)$. This reduces the number from $O^*(n^9)$ to $O^*(n^7)$.

4.3.4.2 An even faster simulation

We use the factor $M = e^{-\lambda cr}$ in the calculation of A to ensure that the result is at most 1. However this factor is smaller than necessary. From the subgradient inequality (4.11), we have

$$f(X_t') \geq f(X_t) + \nabla.(X_t' - X_t),$$

where ∇ is a subgradient at X_t. Now $\nabla.(X'_t - X_t) \sim N(0, \sigma^2 \|\nabla\|^2)$ and $\|\nabla\| \leq \lambda$. Thus, using Lemma 20,

$$\mathbf{P}(\nabla.(X'_t - X_t) > c\lambda r \ln n/\sqrt{n}) < n^{-c^2 \ln n/2}.$$

Hence $F(X'_t)/F(X_t) \leq e^{-c\lambda r \ln n/\sqrt{n}} = M'$, say, with very high probability. Let us write $A' = \sqrt{M'F(X'_t)/F(X_t)}$, $A_0 = \sqrt{M/M'}$, and note that $A_0 > \frac{3}{4}$ is a constant. Then the acceptance probability A equals $A_0 A'$. We write $\mathcal{A} = \mathcal{A}_0 \mathcal{A}'$ for the corresponding indicator variables. Hence we may simulate a step of the original algorithm by first choosing $\mathcal{A}_0$ and generating $\mathcal{A}'$ only if $\mathcal{A}_0 = 1$.

Now consider the algorithm which uses only $\mathcal{A}'$. To each step of the modified algorithm there will be a geometrically distributed number of steps of the original algorithm with expectation $1/A_0$. However, observe that it does not matter if we stop at a random time in the original algorithm provided that it is independent of the random walk and it exceeds the bound on the coupling time (at least with very high probability).

Thus we need not generate the sum of geometrics if the number of steps of the modified algorithm exceeds the coupling time. Note that the saving in the number steps from generating this sum (as a negative binomial) would be only a small constant factor, and we will not consider it here.

Now consider a k-step of the modified algorithm. We use the same notation as in Section 4.3.4.1. We have, with very high probability, $\|S\| \leq cr\sqrt{k}$, and hence $f(X_t + S) - f(X_t) \leq \lambda cr\sqrt{k}$. Thus

$$a \geq \exp(-\tfrac{1}{2}\lambda cr(k \ln n/\sqrt{n} + \sqrt{k})) = \exp(-\tfrac{1}{2}\beta(k \ln n/\sqrt{n} + \sqrt{k})),$$

and hence $(1 - a) \leq \tfrac{1}{2}\beta(k \ln n/\sqrt{n} + \sqrt{k})$. With this modification (4.26) becomes

$$t_c(k) \leq 2 + e^{\tfrac{1}{2}\beta(k \ln n/\sqrt{n} + \sqrt{k})/2} t_c(\tfrac{1}{2}k),$$

from which it follows that

$$t_u(k) \leq 1 + e^{\beta(k \ln n/\sqrt{n} + 2\sqrt{k})} \beta(k \ln n/\sqrt{n} + \sqrt{k}) \log_2 k.$$

Hence we may choose $k = 2^{\lfloor \log_2(\sqrt{n}/\beta \ln n) \rfloor}$, and simulate $\Omega(\sqrt{n}/r)$ steps of the modified algorithm in $O(\log(\sqrt{n}/r))$ k-steps. This reduces the required number of "steps" to $O^*(n^{6.5})$.

4.3.5 Conclusions

We have given a coupling proof for the polynomial-time convergence of a random walk for generating an approximately uniformly distributed point in a convex body. The argument is completely elementary, and does not require any auxiliary proof of an isoperimetric inequality, as with the usual approach. On the other hand, the running time of the algorithm is inferior to the best conductance methods. We obtain running time $O^*(n^{6.5})$, whereas Kannan, Lovász, and Simonovits [75] obtain $O^*(n^5)$ (and an amortized time for multiple points which is even smaller). However, the present conductance techniques are the product of years of refinement. We have drawn on some of these developments, but our proof is a first attempt using a different approach. It would be surprising if it cannot be improved further. The present difficulty is the small step size r. We have described modifications to mitigate this problem, but genuine progress would result from an improvement in the arguments to allow r to be increased.

Intermezzo

Path Coupling

In this chapter we examine the technique known as *path coupling*, introduced by Dyer and the author [21, 27], and subsequently used in a series of papers [23, 25, 31, 43, 45, 46, 90]. A recent extension of the path coupling technique, utilizing ideas similar to those found in Section 4.3.4.1, is due to Czumaj *et al.* [32]. In this chapter, we slightly generalize the original path coupling argument [21, 27] by omitting a symmetry requirement. A similar and independent refinement was recently made by Dyer and Greenhill [47].

The essence of the path coupling methodology is very simple. We saw in Chapter 4 that when applying traditional coupling techniques we need to consider all pairs of states, and show that for all such pairs, there is a tendency for the two coupled copies of the Markov chain to come closer together.

With path coupling, we remove the need to consider *all* pairs of states, by defining a *path*, or sequence of states between an arbitrary pair of states. We then only need to consider pairs of states that are adjacent on some path. Note that states that are adjacent on a path are not necessarily adjacent states in the Markov chain, although they are in the theorems in this chapter.

Suppose we have a path defined between each pair of states, and a coupling that is defined on all pairs of states that are adjacent on some path. Suppose further that at each step, this coupling reduces the expected distance between adjacent pairs (under some appropriate distance quasi-metric[1]). Then by linearity of expectation and the triangle inequality we may conclude that every path is contracting in expectation, thus a simple induction will conclude a proof of rapid-mixing. Alternatively, if we can only show that for each pair of (path-wise) adjacent states the expected distance is non-increasing, then we may conclude that the length of any path as a whole is also non-increasing in expectation; if we then have the further condition that the variance of the length of the entire path is sufficiently large, then we may once again conclude that the Markov chain is rapidly mixing.

Formally, consider a Markov chain, $\mathcal{M}$, with finite state space Ω. Suppose R is a relation on Ω, such that the transitive closure of R is the complete relation on

---

[1] A quasi-metric relaxes the definition of a metric so as not to insist on symmetry.

Ω. Let $\delta : R \to [0, \infty)$ be a function such that $\delta(a, b) = 0 \Rightarrow a = b$; we will call this a *proximity function*. δ defines the distance between "adjacent" pairs, we shall extend this definition to be a distance function on the entire state space. Let $\delta^* : (\Omega \times \Omega) \to [0, \infty)$ be the distance function defined by

$$\delta^*(a, b) = \begin{cases} 0 & \text{if } a = b, \\ \min_{\substack{r_1, r_2, \ldots, r_k \in R \\ 1 \leq k \leq |\Omega| - 2}} \{\sum_{i=1}^{k} \delta(r_i) : r_1 r_2 \ldots r_k = (a, b)\} & \text{otherwise.} \end{cases}$$

Observe that δ^* satisfies the triangle inequality, and $\delta^*(a, a) = 0$ for all a, i.e. δ^* is a quasi-metric. If δ is symmetric, then δ^* is a metric. δ^* will be our distance function: it implicitly defines a shortest path between every pair of states; we will refer to these paths as *geodesic paths*.

We shall define a *special coupling procedure* to be a particular type of coupling procedure: one that need only be defined for pairs in R, and that may be decomposed into the following two operations. Given a current state $(a, b) \in R$, we first pick a', according to the transition matrix of $\mathcal{M}$, and then choose b' from the marginal distribution of (a', b'), given a'. Note that all coupling procedures may be reformulated as special coupling procedures when Ω is finite; we make the distinction merely to emphasize the manner in which the coupling procedure is applied.

Given a special coupling procedure, and a geodesic path between all pairs of states in Ω, we may define an *induced coupling procedure* on the whole of $\Omega \times \Omega$. For an arbitrary pair of states, (a, b), with path $a = a_1, a_2, \ldots, a_{k+1} = b$, we first pick a_1' according to the transition matrix of $\mathcal{M}$, and then choose a_2' from the marginal distribution of (a_1', a_2'), given a_1'. Given a_2', we then choose a_3' from the marginal distribution of (a_2', a_3'), under the special coupling procedure, and so forth.

PATH COUPLING LEMMA (LEMMA 24). *Given a special coupling procedure such that for all $(a, b) \in R$ we have*

$$\mathbf{E}\left(\delta^*(a', b')\right) \leq \beta \delta(a, b)$$

where β is independent of a and b. Then this inequality extends to an arbitrary pair of states under the induced coupling procedure.

Proof. For an arbitrary pair of states, (a, b), let $r_1 = (a_1, a_2), r_2 = (a_2, a_3), \ldots r_k = (a_k, a_{k+1})$ be the path that is used in the definition of $\delta^*(a, b)$ (so $a = a_1$ and $b = a_{k+1}$). Then under the induced coupling procedure we have

$$\mathbf{E}\left(\delta^*(a', b')\right) \leq \mathbf{E}\left(\sum_{i=1}^{k} \delta^*(a_i', a_{i+1}')\right) \qquad \text{(by minimality of } \delta^*\text{)}$$

$$\leq \sum_{i=1}^{k} \mathbf{E}\left(\delta^*(a_i', a_{i+1}')\right) \qquad \text{(by linearity of expectation)}$$

$$\leq \sum_{i=1}^{k} \beta \delta(a_i, a_{i+1}) \qquad \text{(by definition of } \beta\text{)}$$

$$= \beta \delta^*(a, b). \qquad \text{(by definition of } \delta^*\text{)}$$

$\square$

If $\beta < 1$, then it follows that $\mathcal{M}$ is rapidly mixing. Specifically, if $D = D(\mathcal{M}, \delta^*)$ is the maximum value of δ^*, and $d = d(\mathcal{M}, \delta^*)$ is the minimum strictly positive value of δ^*, then the expected distance between two points evolving under the induced coupling procedure is at most $\beta^t D$, and thus the probability that coupling has not occurred is at most $\beta^t D / d$.

If, instead, $\beta \leq 1$, and we can show for any pair of states that under the induced coupling procedure there is some polynomial number of steps in which there is a constant probability of the distance changing, then we once again have rapid-mixing.

Remark. It should be clear that the path coupling methodology can also be applied when the state space is not finite, although there will then be additional technical considerations to deal with, resulting from the replacement of the minimums in the definition of δ^* and d with infimums, and the fact that reformulating a coupling procedure as a special coupling procedure may not be possible on some set of zero measure. $\bigcirc$

We can apply the path coupling method in a quite general setting for particular Markov chains on product spaces. Before we do this we shall need to set up some notation.

Let V and C be finite sets, and define $n = |V|$ and $k = |C|$. For our theorems, we typically consider a finite Markov chain $\mathcal{M}$, with state space $\Omega \subseteq C^V$, the set of functions from V to C, and unique equilibrium distribution π. The reader may find it helpful to keep in mind the example of *proper graph colourings* (we analyse this example in depth in Section 5.8): then V is the set of vertices of a graph, and C is a set of colours; the Markov chain $\mathcal{M}$ then has as state space the set of functions from vertices to colours, and as equilibrium distribution the uniform distribution on the set of *proper colourings*. In our first theorem, we require that $\Omega = C^V$, but in our second theorem we do not make this assumption.

For $X \in \Omega$, $v \in V$, and $c \in C$, let us use the notation $X_{v \to c}$ to denote the state resulting from making the transition at X associated with the pair (v, c). Thus

$$X_{v \to c}(w) = \begin{cases} c & \text{if } w = v, \text{ and} \\ X(w) & \text{otherwise.} \end{cases}$$

Using this notation, we may more precisely define the transition structure of $\mathcal{M}$. We first pick $v \in V$ from a fixed distribution J on V. Then we pick $c \in C$ according to a distribution $\nu_{X,v}$ on C, dependent only on the current state X and v, and make the transition to $X_{v \to c}$. We assume that $X_{v \to c} \notin \Omega$ implies that $\nu_{X,v}(c) = 0$.

In the two theorems in this chapter, we consider a class of Markov chains in which the (not necessarily unique) choice of path arises naturally: we consider Markov chains with state space (some subset of) the set of functions from V to C. Our paths will be constructed simply by insisting that adjacent states on a path differ in their mapping of at most one $v \in V$. The statement of the theorems may seem technical, but their proofs are quite elementary.

The distance function that we shall use in the first theorem is Hamming distance, which for states X and Y we shall denote as $H(X, Y)$, i.e. $H(X, Y)$ is simply the number of $v \in V$ such that $X(v) \neq Y(v)$. Thus adjacent states on a path have unit Hamming distance.

In both of the theorems, β will be an upper bound on the expected distance between adjacent states after a single time-step.

GENERAL PATH COUPLING THEOREM (THEOREM 4). *Let* $\Omega = C^V$, *and*

$$\beta = \max_{X,Y \in \Omega, i \in V} \left\{ 1 - J(i) + \sum_{j \in V} J(j) d_{\mathrm{TV}}(\nu_{X,j}, \nu_{Y,j}) \,\middle|\, \right.$$
$$\left. Y = X_{i \to c} \text{ for some } c \in C, \text{ and } Y \neq X \right\}.$$

If $\beta < 1$, *then the mixing rate of* $\mathcal{M}$ *satisfies* $\tau(\varepsilon) \leq \lceil \ln(n\varepsilon^{-1})/\ln \beta^{-1} \rceil$, *and the relaxation time satisfies* $\tau_2 \leq 1/(1 - \beta)$.

Proof. Suppose σ_1 and σ_2 are distinct probability distributions on C. Then define the probability distribution $(\sigma_1 - \sigma_2)^+$ by

$$(\sigma_1 - \sigma_2)^+(c) = \frac{\max\{0, \sigma_1(c) - \sigma_2(c)\}}{d_{\mathrm{TV}}(\sigma_1, \sigma_2)}.$$

Define the special coupling for $\mathcal{M}$ at state (X, Y) (where $H(X, Y) = 1$) by selecting the next state (X', Y') according to the following experiment.

1. Choose $v \in V$ according to J and $c_X \in C$ according to $\nu_{X,j}$.

2. With probability $\nu_{Y,j}(c_X)/\nu_{X,j}(c_X)$, let $c_Y = c_X$, otherwise pick c_Y according to $(\nu_{Y,j} - \nu_{X,j})^+$.

3. Make the transition to (X', Y'), where $X' = X_{v \to c_X}$, and $Y' = Y_{v \to c_Y}$.

Observe that, marginally, we choose c_Y according to $\nu_{Y,j}$, and that this is the "maximal coupling" (i.e. best possible coupling) between $\nu_{X,j}$ and $\nu_{Y,j}$. In particular $\mathbf{P}(c_Y \neq c_X) = d_{\mathrm{TV}}(\nu_{Y,j}, \nu_{X,j})$.

Suppose, in the special coupling procedure, that X and Y differ only at i. Then

$$\begin{aligned}
\mathbf{E}(H(Y', X')) &= 1 - \mathbf{P}(H(Y', X') = 0) \\
&\quad + \mathbf{P}(H(Y', X') = 2) \\
&= 1 - J(i)\mathbf{P}(c_Y = c_X \mid v = i) \\
&\quad + \sum_{j \neq i} J(j)\mathbf{P}(c_Y \neq c_X \mid v = j) \\
&= 1 - J(i)(1 - d_{\mathrm{TV}}(\nu_{Y,i}, \nu_{X,i})) \\
&\quad + \sum_{j \neq i} J(j) d_{\mathrm{TV}}(\nu_{Y,j}, \nu_{X,j}) \\
&\leq \beta.
\end{aligned}$$

Thus, applying the Path Coupling Lemma, we have that for an arbitrary pair (X, Y), under the induced coupling, $\mathbf{E}(H(X', Y')) \leq \beta H(X, Y)$.

Thus, by iterating this inequality, if X_t and Y_t are the positions of X and Y after t steps, we have that $\mathbf{E}\left(H(X_t, Y_t)\right) \leq \beta^t n$. Furthermore, since H is a non-negative integer-valued function, $\mathbf{P}(X_t \neq Y_t) \leq \beta^t n$. Applying the Coupling Lemma, we see that $d_{\mathrm{TV}}(\mu_t, \pi) \leq \beta^t n$. Taking logarithms and rearranging establishes the mixing rate in the theorem. The relaxation time then follows from Lemma 1. $\qquad\square$

Remark. Suppose we have only that $\beta \leq 1$: then the above theorem appears to tell us nothing about the convergence of the chain. However, the same proof shows that at each step, $H(X, Y)$ cannot increase in expectation, and its value can change either by zero or by one at each step. Suppose that the probability of its value changing at each step is bounded below by α. Then the expected time for the processes to couple, i.e. for $H(X, Y)$ to reach zero, is bounded above by the expected time for a symmetric random walk on the integers $\{0, 1, \ldots, n\}$, started at n and with probability α of moving to an adjacent integer, to reach zero. This is $\alpha^{-1}(n^2 + n)/2$. Using Markov's inequality, we see that the probability that X and Y have not coupled by time t is bounded above by $\alpha^{-1}(n^2 + n)/2(t + 1)$. In particular, in order to ensure that the probability that we have not coupled is no greater than e^{-1}, it suffices to simulate $\tau = \lceil e\alpha^{-1}(n^2 + n)/2 \rceil - 1$ steps of $\mathcal{M}$. Since we may run successive, independent coupling "trials" of length τ, in order to ensure the probability that we have not coupled is bounded above by ε, it suffices to simulate $\mathcal{M}$ for $\lceil \ln\left(\varepsilon^{-1}\right) \rceil \tau$ steps. Thus we will have rapid-mixing whenever we can show that α^{-1} is polynomial in n and c. We will not give general conditions here for this to be true, but we will consider this observation below. $\qquad\bigcirc$

The problem with Theorem 4 is that it requires $\Omega = C^V$. We will relax this assumption by considering a slightly different coupling, for the particularly important class of "Metropolis" Markov chains. The transitions here are as follows. Choose the desired stationary distribution, π. Pick $v \in V$, as before, according to some fixed distribution J. Pick $c \in C$ uniformly at random. Then, with probability $A_{X,v}(c) = \min\{1, \pi(X_{v\to c})/\pi(X)\}$, *accept* and make the transition to $X_{v\to c}$; otherwise *reject* and remain at X. This procedure determines the distributions $\nu_{X,v}$.

The metric that we use in the following theorem is slightly different from the first. In this theorem, we will use the *minimum transition distance*, which we shall denote as $\delta^*(X, Y)$. This is simply the minimum number of transitions of the Markov chain that could be performed in order to move from X to Y. Thus, here D is the diameter of the Markov chain.

METROPOLIS PATH COUPLING THEOREM (THEOREM 5). *Let* $\Omega \subseteq C^V$. *For states* X *and* Y *in* Ω, *such that* $Y \neq X$ *and* $Y = X_{i\to c}$, *let*

$$S(X, Y) = \frac{1}{k} \sum_{c \in C} \min\{A_{X,i}(c), A_{Y,i}(c)\}$$

$$F_j(X, Y) = \frac{1}{k} \sum_{c \in C} |A_{X,j}(c) - A_{Y,j}(c)| \quad (j \neq i).$$

Also define

$$\beta = \max_{X,Y \in \Omega, i \in V} \left\{ \left| 1 - J(i)S(X,Y) + \sum_{j \neq i} J(j)F_j(X,Y) \right| \right.$$

$$\left. Y = X_{i \to c} \text{ for some } c \in C \text{ such that } Y \neq X \right\},$$

$$\eta = \min \left\{ \min_{X \in \Omega, i \in V, c \in C} \{ A_{X,i}(c) \mid A_{X,i}(c) > 0 \}, \right.$$

$$\left. \min_{X,Y \in \Omega, i \in V, c \in C} \{ A_{X,i}(c) - A_{Y,i}(c) \mid A_{X,i}(c) > A_{Y,i}(c) \} \right\}.$$

1. *If $\beta < 1$, then the mixing rate satisfies $\tau(\varepsilon) \leq \lceil \ln(D\varepsilon^{-1})/\ln \beta^{-1} \rceil$, and the relaxation time satisfies $\tau \leq 1/(1 - \beta)$;*

2. *If $\beta \leq 1$, then then the mixing rate satisfies*

$$\tau(\varepsilon) \leq \lceil \ln(\varepsilon^{-1}) \rceil \lceil e\eta^{-1}k(D^2 + D)/\min_{i \in V}\{J(i)\} - 1 \rceil.$$

Remark. Note that if all acceptance probabilities are either 0 or 1 (e.g. if π is the uniform distribution), then $\eta = 1$. Furthermore, if π is the uniform distribution and $k = 2$, then for all X and Y in the theorem, $S(X,Y) = 1$. $\bigcirc$

Proof. This proof is very similar to that of the General Path Coupling Theorem, although we use a different coupling here.

For two states X and $Y \in \Omega$, let $\delta^*(X,Y)$ be the minimum number of transitions required to move from X to Y. Observe that δ^* is a metric. We will let $\delta^* = \delta^*(X,Y)$. Let $X = Z_0, Z_1, \ldots, Z_{\delta^*}$ be such a minimal sequence of transitions, and note that all these states are distinct.

Suppose we have $(\delta^* + 1)$ instances of $\mathcal{M}$, with current states $Z_0, Z_1, \ldots, Z_{\delta^*}$, which will evolve jointly by the following experiment:

1. Choose $v \in V$ according to J and $c \in C$ uniformly at random.

2. Choose W uniformly from $[0,1]$. For $a = 0, 1, \ldots, \delta^*$, if $A_{Z_a,v}(c) \geq W$, accept and move from Z_a to $(Z_a)_{v \to c}$, otherwise reject and remain at Z_a.

This defines a coupling on X and Y. We use Z'_a to denote the state moved to from Z_a in the above experiment.

Observe, by the assumption on transitions of $\mathcal{M}$, that Z_{a-1} and Z_a differ for exactly one element of V, say i. Then

$$\mathbf{E}\left(\delta^*\left(Z'_{a-1}, Z'_a\right)\right) = 1 + \mathbf{P}\left(\delta^*\left(Z'_{a-1}, Z'_a\right) = 2\right) - \mathbf{P}\left(\delta^*\left(Z'_{a-1}, Z'_a\right) = 0\right),$$

since the coupling procedure ensures that $\delta^*\left(Z'_{a-1}, Z'_a\right)$ takes only values in $\{0, 1, 2\}$.

Now, $\delta^* \left(Z'_{a-1}, Z'_a \right) = 0$ only if we choose $v = i$ in the coupling procedure, and both Z_{a-1} and Z_a accept. Thus we have that $\mathbf{P} \left(\delta^* \left(Z'_{a-1}, Z'_a \right) = 0 \right) = J(i) S \left(Z_{a-1}, Z_a \right)$.

The event $\delta^* \left(Z'_{a-1}, Z'_a \right) = 2$ can occur only if we choose $v \in V$ with $v \neq i$, and we accept for precisely one of Z_{a-1} and Z_a. Thus $\mathbf{P} \left(\delta^* \left(Z'_{a-1}, Z'_a \right) = 2 \right) = \sum_{j \neq i} J(j) F \left(Z_{a-1}, Z_a \right)$.

If we apply the Path Coupling Lemma, it follows that $\mathbf{E} \left(\delta^* \left(X_t, Y_t \right) \right) \leq \beta^t D$, and since δ^* is a non-negative integer-valued function, we thus have $\mathbf{P} \left(X_t \neq Y_t \right) \leq \beta^t D$. Applying the Coupling Lemma, we see that the variation distance from equilibrium after t steps is bounded above by $\beta^t D$. Taking logarithms and rearranging establishes the mixing rate in the theorem. The relaxation time then follows from Lemma 1.

To establish the second part of the theorem, we will assume that all acceptance probabilities $A_{X,v}(c) \leq 1/2$ in $\mathcal{M}$. If this is not the case, we may simply halve all the acceptance probabilities to make it so. This is equivalent to having a "do nothing with probability $1/2$" condition at the beginning of each step, and at most doubles the expected number of steps for $\mathcal{M}$ to couple. We allow for this in our calculations. Note that if $\beta \leq 1$ originally, this will still hold in the revised chain.

Consider the sequence of values taken by $\delta^*(X, Y)$ under the coupling. Assuming $\beta \leq 1$, then at each step, $\delta^*(X, Y)$ cannot increase in expectation, and its value may change either by zero or by one. Suppose that the probability that its actual value changes is bounded below by a. Then the expected time for the processes to couple is bounded above by the expected time for a symmetric random walk on the integers $\{0, 1, \ldots, D\}$, with probability a of moving to an adjacent state, started at D, to reach zero. This is $a^{-1}(D^2 + D)/2$.

Using Markov's inequality, we see that the probability that we have not coupled by time t is bounded above by $a^{-1}(D^2 + D)/2(t+1)$. In particular, in order to ensure that the probability we have not coupled is no greater than e^{-1}, it suffices to simulate $\tau = \lceil ea^{-1}(D^2 + D)/2 \rceil - 1$ steps of $\mathcal{M}$. Since we may run successive, independent coupling "trials" of length τ, in order to ensure that the probability that we have not coupled is bounded above by ε, it suffices to simulate $\mathcal{M}$ for $\lceil \ln \left(\varepsilon^{-1} \right) \rceil \tau$ steps.

It remains to show, however, that $a \geq \eta \min_{i \in V} \{ J(i) \}/k$. We do not do this for the coupling above, but for one which is subtly different. Suppose there is a pair of states, X and Y, such that any choice of $v \in V$, $c \in C$, $W \in [0, 1]$ in the previous coupling would result in $\delta^*(X, Y)$ remaining unchanged: we will call this a "stuck pair". If there is no such pair then $\delta^*(X', Y') \neq \delta^*(X, Y)$ with probability at least $\eta \min_{i \in V} \{ J(i) \}/k$, and we are done.

Let i be the element of V that is different in X and Z_1, and let $c_1 = Z_1(i)$, so $X(i) \neq c_1$. Also $c_1 \neq Y(i)$, otherwise choosing i and $Y(i)$ would reduce $\delta^*(X, Y)$ with probability at least η. Let $A, A' \leq 1/2$ be the acceptance probabilities for $Z_1(i)$ in X and Y, given $v = i$. Clearly $A > 0$. If $A > A'$ then we can have $X' = Z_1, Y' = Y$ with probability at least η and we are done. Also if $A < A'$, we can have $X' = X, Y' = Y_{i \rightarrow c_1} = Y^*$, say, with probability at least η, and hence we must have $\delta^*(X, Y^*) = \delta^*(X, Y)$. If $A = A'$, then we only have $\delta^*(X, Y^*) \leq \delta^*(X, Y) + 1$.

Our coupling will then be modified as follows. If X and Y are not stuck or $v \neq i$ or $c \neq c_1$, use the previous coupling. Otherwise, accept in X if $W \leq A$ and in Y if

$(1 - W) \leq A'$. Note that X and Y are still faithful copies of $\mathcal{M}$. Now, conditional on i and c_1, we have

$$(X', Y') = \begin{cases} (Z_1, Y) & \text{with probability } A, \\ (X, Y) & \text{with probability } (1 - A - A'), \\ (X, Y^*) & \text{with probability } A'. \end{cases}$$

Thus $\mathbf{P}\left(\delta^*\left(X', Y'\right) \neq \delta^*(X, Y)\right) \geq A \geq \eta$. If $A < A'$, then $\mathbf{E}\left(\delta^*\left(X', Y'\right)\right) - \delta^*(X, Y) = -A < 0$. If $A = A'$, then $\mathbf{E}\left(\delta^*\left(X', Y'\right)\right) - \delta^*(X, Y) \leq -A + A' = 0$. So, in either case, we preserve the condition that δ^* does not increase in expectation. But the probability that it changes is now at least $\eta \min_{i \in V}\{J(i)\}/k$, as required.

Applying the Coupling Lemma completes the proof. $\qquad\square$

It should be noted the definitions of β in the two path coupling theorems may yield different values when maximized over adjacent states $X, Y \in \Omega$. In general the first is a lower bound on the second, and this inequality can be strict. They do however coincide in the case $k = 2$, as may be verified by easy calculations.

Chapter 5

Applications: Path Coupling

5.1 Introduction

In this chapter we consider a variety of applications of the path coupling method to show rapid-mixing (or in some cases more rapid-mixing) of certain Markov chains. In each case, the associated exact counting problem is #P-complete.

In Section 5.2, we rework our analysis of sink-free orientations from Section 4.2, using path coupling and obtain a tighter bound on the mixing time with a far simpler proof.

In Section 5.3 we consider the related problem of sampling from the set of sink- and source-free orientations of a graph.

In Section 5.4 we give a Markov chain with stationary distribution the set of totally edge cyclic orientations of a graph, and show that if this Markov chain is irreducible, then it is also rapidly mixing.

In Sections 5.5, 5.6, and 5.9 we consider several problems from statistical physics: the hard-core model, both conserved and non-conserved, and the anti-ferromagnetic Potts model.

In Section 5.7 we consider the problem of sampling uniformly from the set of linear extensions of a partial order, and provide the first tight (up to a multiplicative constant) bounds on the mixing and relaxation times for an elementary Markov chain on a state space that is #P-complete to count.

Finally, in Sections 5.8 and 5.10, we consider the problem of graph k-colouring, a problem which we looked at in the context of hypergraphs in Section 4.1.

5.2 Twice-Sat Revisited

Recall the problem of Twice-Sat [20], which we considered in Section 4.2. We may cast this problem into a form in which we may apply the Metropolis Path Coupling Theorem, by considering satisfying assignments to variables in the Boolean formula to be members of the set $\Omega \subseteq \{\text{True}, \text{False}\}^V$, where V is the set of variables.

We consider the same Markov chain (up to isomorphism) as we did in Section 4.2, i.e. we let $\mathcal{M}$ be the single-site Metropolis Markov chain on Ω. Transitions on $\mathcal{M}$ are thus made by choosing $v \in V$ and truth value c uniformly at random. If assigning c to v results in a satisfying assignment, we accept and make a transition to this state.

We have from Section 4.2 that under (non-restrictive) conditions, $\mathcal{M}$ is ergodic.

Let m be the number of clauses (and $n = |V|$ the number of variables)[1]. Corollary 10 then gives us that D, the diameter of $\mathcal{M}$, satisfies $D \le \max\{2m+1, n\}$.

We may now apply Theorem 5, the Metropolis Path Coupling Theorem to this chain. If X and Y are two assignments that differ only at i, we shall write $Y = X_{\bar{i}}$.

Thus, using the notation of Theorem 5, we see that (for $j \ne i$),

$$F_j(X, X_{\bar{i}}) = \frac{1}{2}\left[|A_{X_{\bar{i}},j}(\text{True}) - A_{X_{\bar{i}},j}(\text{True})| + |A_{X_{\bar{i}},j}(\text{False}) - A_{X_{\bar{i}},j}(\text{False})|\right]$$

$$= \begin{cases} 1 & \text{if precisely one of } X_{\bar{j}}, Y_{\bar{j}} \in \Omega, \\ 0 & \text{otherwise.} \end{cases}$$

Thus, we may write down a formula for β:

$$\beta = 1 - 1/n + \max_{X \in \Omega; i \in V} |\{j \in V \mid j \ne i; \text{precisely one of } X_{\bar{j}} \text{ and } Y_{\bar{j}} \in \Omega\}|/2n,$$

where $Y = X_{\bar{i}}$.

Now if only one of $X_{\bar{j}}$ and $Y_{\bar{j}}$ is in Ω, this must be caused by a clause in which both i and j appear as variables. Furthermore, in this clause, the literal of variable j must appear as True, and the literal of i must appear as False in one of X and Y and True in the other. All other literals in the clause must appear as False. Clearly there can be at most one such j in each clause in which i appears, i.e. there are at most two such js. Thus $\beta \le 1$.

Applying conclusion 2 of the Metropolis Path Coupling Theorem (with $\eta = 1$), we see immediately that $\mathcal{M}$ is rapidly mixing with mixing rate $O(n^3 \log \varepsilon^{-1})$. This compares favourably with our rapid-mixing result using traditional coupling (Lemma 14), which yielded a mixing rate of $O((n^3 + nm^3) \log \varepsilon^{-1})$, and required a much lengthier analysis.

Using the reduction from approximate counting to approximate sampling which was explained in Section 4.2.5, it follows that there is an *fpras* for Ω that runs in time $O(m^2 n^3 \varepsilon^{-2} \log n\varepsilon^{-1})$.

5.3 Sink- and Source-Free Graph Orientations

5.3.1 Introduction

In Sections 4.2 and 5.2 we looked at sink-free graph orientations. A natural allied problem is thus looking at graph orientations with neither source nor sink. A sink-

---

[1]Note that the definitions of m and n have been swapped from Section 4.2, so as to agree with the notation of the Intermezzo.

and source-free orientation ("SSFO") of a graph is an assignment of a direction to each of the edges of a graph in such a fashion that every vertex has positive in-degree and out-degree.

In this section, we consider the standard combinatorial problems associated with the set of SSFOs of a graph. Specifically, we show that decision and construction may be accomplished in linear time, illustrate a polynomial delay listing algorithm, show that exact counting is #P-complete, and show how sampling and approximate counting may be accomplished for all but a small class of graphs.

The problem of approximately sampling from the set of SSFOs of a graph was considered briefly by Dyer and the author [21]. We recapitulate that result here and extend it to a larger class of graphs.

Let $G = (V, E)$ be an undirected (multi)graph with $n = |V|$ and $m = |E|$. Let $\Omega = \Omega(G)$ be the set of SSFOs of G.

When we consider algorithmic issues we assume that G is represented as an adjacency-list structure.

5.3.2 Decision and Construction

In this section we show that the problems of deciding SSFO, and constructing a witness if one exists, are computationally simple tasks. Specifically, for a graph with n vertices and m edges, we may accomplish the tasks of construction and decision in time $O(m + n)$. To show this, we first provide a lemma characterizing graphs which admit an SSFO.

LEMMA 25. *A graph, G, has a sink- and source-free orientation if and only if it has minimum degree 2.*

Proof. If a graph admits an SSFO, it clearly has minimum degree 2. We will establish the converse by induction on the number of vertices of the graph.

Given a graph G, with degree at least 2, and a non-loop edge of it, $e = \{a, b\}$, we have by our induction hypothesis an SSFO of $G \cdot e$, (since contracting an edge cannot decrease the minimum degree of a graph). This orientation of $G \cdot e$ induces an orientation of the edges of G, save e. In this induced partial orientation, there must be edges oriented both towards and away from the set $\{a, b\}$. Since both a and b have degree at least 2, there must be an edge oriented into one (without loss of generality, a), and an edge oriented out of the other. Thus orienting e from a to b ensures that we have an orientation of G that is sink- and source-free, which establishes the induction step. The base cases, in which all edges are loops, are trivial. $\square$

COROLLARY 12. *Deciding SSFO, and constructing a witness if one exists, may be accomplished in time linear in the number of vertices and edges of the graph.*

Proof. The induction explained in the proof of Lemma 25 may readily be implemented as a recursive construction. $\square$

5.3.3 Exact Counting

5.3.3.1 *Self-reducibility*

The deletion of an edge in an SSFO, O, induces a unique orientation $O - e$ in the resulting graph $G - e$ which may not be an SSFO. However, the contraction of an edge in an SSFO always induces an SSFO, $O \cdot e$, of the resulting graph.

LEMMA 26. *Suppose $G = (V, E)$ is a graph that admits an SSFO, and $e \in E$ is a non-loop edge. Then there is a natural bijection between $\Omega(G)$ and $\Omega(G-e) \cup \Omega(G \cdot e)$.*

Proof. Let $<$ denote an arbitrary total order on V. We will denote the end-points of e by a and b, and without loss of generality, assume $a < b$. Define $F : \Omega(G) \to \Omega(G - e) \cup \Omega(G \cdot e)$ in the following way: if O is an SSFO of G in which e points towards a, and the orientation obtained from O by reversing the orientation of e is also an SSFO, then $F(O) = O - e \in \Omega(G - e)$. Otherwise, $F(O) = O \cdot e \in \Omega(G \cdot e)$. To see that this function does generate orientations in the range claimed, consider the two cases used in the construction of F — in the first case, there must exist edges oriented both towards and away from both a and b in O, hence $O - e \in \Omega(G - e)$; in the second case, there must exist at least one edge (other than e) oriented towards a or b, and at least one edge (again, other than e) oriented away from a or b. Thus $O \cdot e \in \Omega(G \cdot e)$. Thus F is injective. Conversely, given a sink- and source-free orientation of either $G \cdot e$ or $G - e$, it is clear that these may be extended to an orientation in G that maps back to our original orientation of $G \cdot e$ or $G - e$ respectively, i.e. F is surjective too. $\square$

COROLLARY 13. $N_0(G) = N_0(G - e) + N_0(G \cdot e)$ *and* $N_0(G \cdot e) \geq N_0(G - e)$.

Proof. The first claim follows immediately from the lemma. The second follows from consideration of the map F in the proof of the lemma. $\square$

COROLLARY 14. *SSFO is a self-reducible decision problem.*

Proof. This is immediate from Lemma 26 and the definition of self-reducibility [108]. $\square$

COROLLARY 15. *Listing all sink- and source-free orientations of a graph has a polynomial space, polynomial delay [72] algorithm.*

Proof. We may use the recurrence of Lemma 26 as the basis for a recursive algorithm which has as base cases those graphs which have no SSFO (which we know from Lemma 25) and graphs in which all edges are loops. The base cases are elementary. Provided we ensure that the edge chosen for the recursion is never a loop, and that the recursion is evaluated lazily, this algorithm is clearly polynomial space. To see that it has polynomial delay, consider the recursion tree as a search tree in the algorithm. Appealing once more to Lemma 25 we see that in traversing from one SSFO to the next, we are making only $O(m)$ moves on the search tree. $\square$

Remark. Valiant [125] notes the existence of a similar recursive argument, which, as Goldberg [55] points out, extends to any self-reducible structure. $\bigcirc$

Figure 5.1: Two graphs with identical Tutte polynomials.

There is a striking similarity between the contraction-deletion relationship of Lemma 26 and that defining the *Tutte polynomial* (see Section 1.4.1). However, the definition of the base cases does not match, so SSFO does not appear to represent any point in the "Tutte plane". Indeed we can prove a stronger statement than this:

PROPOSITION 11. *#SSFO is not a function of the Tutte polynomial.*

Proof. Recall the recursive definition of the Tutte polynomial (see Section 1.4.1) which defines, for a loop e, the following reduction: $T(G; x, y) = yT(G - e; x, y)$.

It is clear from this reduction that the two graphs in Figure 5.1 share the same Tutte polynomial[2], since we may delete a loop from each, so as to obtain the same graph. By inspection, however, it can be seen that the graph on the right admits 32 SSFOs, whereas the graph on the left admits only 16. $\square$

5.3.3.2 #P-completeness of #SSFO

We next turn to a significant, but not unexpected result: that counting SSFOs exactly is a #P-complete problem. We do this by reducing the problem of counting sink-free graph orientations to it; recall we showed in Section 4.2.3 that counting sink-free graph orientations is #P-complete.

We proceed by showing a non-parsimonious reduction from #SFO: we will show that if we can count SSFOs, then we can calculate the number of orientations with no sinks, and a specified number of sources—the result then follows almost immediately.

THEOREM 6. *Counting the number of sink- and source-free orientations of a graph is #P-complete.*

Proof. We will proceed by reducing the problem of counting sink-free graph orientations to counting SSFOs. Since counting sink-free orientations is a #P-complete problem (Theorem 3) this will establish the claimed result.

Suppose initially that we wish to count the number of sink-free orientations of a graph G. Define $N_{i,j}$ to be the number of orientations of G with i sources and j sinks. The number of sink-free graph orientations of G is then $\sum_{r=0}^{n} N_{r,0}$.

Let us define G_A to be the graph formed by adding a vertex, v, to G, and adding edges between all vertices of G and v. Let us further define G_B to be the graph formed by adding a self-loop to v in G_A (or equivalently, adding vertices v_1 and v_2 and edges so as to form a triangle from v, v_1, and v_2—it will make no difference to the subsequent analysis).

---

[2]We reprint Figure 4.2 here as Figure 5.1 for convenience.

Table 5.1: Summary of orientations of G_A and G_B.

	G_A	G_B	$G_B - 2G_A$
$N_{0,0}$	$2^n - 2$	2^{n+1}	4
$N_{0,i}$ and $N_{i,0}$	$2^{n-i} - 1$	2^{n-i+1}	2
$N_{i,j}$	2^{n-i-j}	$2^{n-i-j+1}$	0

There are three cases, summarized in Table 5.1. The set of orientations with no
sources and no sinks (of size $N_{0,0}$) can be extended to sink- and source-free orientations of G_A in $2^n - 2$ ways, and to G_B in 2^{n+1} ways. For positive i, the set of
orientations with no sources and i sinks or i sources and no sinks (of size $N_{0,i} + N_{i,0}$)
can be extended to sink- and source-free orientations of G_A in $2^{n-i} - 1$ ways, and to
G_B in 2^{n-i+1} ways. Finally, for positive i and j, the set of orientations with i sources
and j sinks (of size $N_{i,j}$) can be extended to sink- and source-free orientations of G_A
in 2^{n-i-j} ways, and to G_B in $2^{n-i-j+1}$ ways.

Thus the number of sink- and source-free orientations of G_B, less twice the number of sink- and source-free orientations of G_A, is precisely four times the number of
sink-free orientations of G.   □

5.3.4 Approximate Counting and Sampling

In order to sample from the set of SSFOs, one might naturally think of sampling
from the set of sink-free orientations and discarding those which have a source. Unfortunately, this simple form of Monte Carlo sampling will not in general work, as
consideration of the following family of graphs will show.

Let G_n be the graph constructed by taking two copies of the star graph on $n +
1$ vertices, and identifying each pair of vertices except the central vertices (i.e. the
complete bipartite graph $K_{2,n}$). Then G_n has $2^n - 1$ sink- and source-free orientations,
and $3^n - 2^{n+1} + 1$ sink-free orientations; clearly the ratio of SSFOs to sink-free
orientations decreases exponentially as n increases.

Another obvious attempt to make would be to use the same Markov chain as for
sink-free graph orientations, and apply path coupling. This approach is mentioned
briefly in Bubley and Dyer [21]. The approach works to a certain extent, but only for
graphs of minimum degree 4. We recapitulate this method next.

Let $\mathcal{M} = \mathcal{M}(G)$ be the single-site update Metropolis Markov chain on the set
of SSFOs of G. Thus $\mathcal{M}$ has transitions modelled by: pick an edge in G, and an

orientation for that edge, uniformly at random; if the resulting orientation is an SSFO, accept this new orientation as the new state of $\mathcal{M}$, otherwise reject, and remain at the original orientation.

PROPOSITION 12. *If G has minimum degree* 3, *the diameter of $\mathcal{M}$ is bounded above by $n + m$.*

Proof. Let X and Y be two arbitrary SSFOs—we shall show that there are two sequences of transitions that may be made to X and Y respectively, which will transform these two SSFOs to the same SSFO, and that the total number of transitions is at most $n + m$. Proceed as follows: consider D, the set of edges that have different orientations in X and Y. If any edge in D can be flipped in either X or Y, so as to maintain an SSFO then perform the flip, and hence remove the edge from D. Repeat until there is no edge in D that can be flipped in either X or Y.

We will now consider Y to be fixed, and show that we can perform transitions on X to bring it to Y. First some more terminology. An edge that cannot be flipped in X will be said to be *critical*; more specifically, if flipping it would result in a sink, it will be said to be *sink-critical*, if it would result in a source, *source-critical*, and if in both, then *sink- and source-critical*. A vertex that could become a sink or a source by the flipping of one edge will be said to be *sink-critical* (or *source critical*, respectively) too. Note that a vertex cannot be both sink- and source-critical, by our assumption that G has minimum degree at least 3. The vertex (or vertices) that will become sources or sinks by the flipping of an edge will be said to be *critical for that edge*.

Consider the structure of D. Let e be an edge in D, and suppose, without loss of generality, that e is sink-critical in X, and that v is the vertex that is critical for e (in X). Since e is of the opposite orientation in Y, there must be at least one other edge incident to v that is in D (or else v would be a sink in Y). In fact there cannot be more than one other edge incident to v that is in D: suppose there were, and that one of them was e'. Then since e' is in D it must be critical in Y. The vertex for which it is critical cannot, by our assumptions, be v—thus it must be source-critical for its other end-point, v', in Y. Similarly e' must be sink-critical for v' in X. This in turn would require each of the other edges incident to v' to be in D, and sink-critical in X, and thus so must all the edges adjacent to these, and so forth. Since a vertex can be sink-critical for only one edge, and we have a finite number of edges, we have established a contradiction. D must therefore consist solely of disjoint cycles—and for each such cycle, every edge incident to the cycle must be oriented in the same fashion: either towards or away from the cycle.

For each disjoint cycle, we shall flip the minimum number of edges necessary to change the configuration such that the cycle has one edge incident to the cycle which is oriented differently from the others. This will allow us to flip the entire cycle, and we can then flip back the edges we had to flip to change the cycle.

So, to summarize, we need to flip each edge at most once to change it from being in D, plus the number of edges we have to flip (twice) to allow us to flip these disjoint cycles in D. Since there can be at most $n/2$ of these disjoint cycles, we see that the total number of transitions is bounded above by $m + n$, the m coming from the fact that every edge may need to be flipped once if initially all edges have different orientations in X and Y, and the n coming from $2 \times n/2$, the number of additional transitions needed to allow us to flip the disjoint cycles. $\square$

We will proceed by applying the Metropolis Path Coupling Theorem (Theorem 5). If X and Y are two SSFOs that differ only in their orientation of edge i, we shall write $Y = X_{\bar{i}}$. Using the notation of Theorem 5, we see that (for $j \neq i$)

$$F_j(X, X_{\bar{i}})$$

$$= \frac{1}{2} \left[|A_{X,j}(\text{Agree}) - A_{X_{\bar{i}},j}(\text{Agree})| + |A_{X,j}(\text{Disagree}) - A_{X_{\bar{i}},j}(\text{Disagree})| \right]$$

$$= \begin{cases} 1 & \text{if precisely one of } X_{\bar{j}}, Y_{\bar{j}} \in \Omega, \\ 0 & \text{otherwise.} \end{cases}$$

Thus, we may write down a formula for β:

$$\beta = 1 - 1/n + \max_{X \in \Omega; i \in V} |\{j \in V \mid j \neq i; \text{ precisely one of } X_{\bar{j}} \text{ and } Y_{\bar{j}} \in \Omega\}|/2n,$$

where $Y = X_{\bar{i}}$.

Let us consider how large $|\{j \in V \mid j \neq i; \text{ precisely one of } X_{\bar{j}} \text{ and } Y_{\bar{j}} \in \Omega\}|$ may be in the maximum. Suppose that X and $Y \in \Omega$, and only one of $X_{\bar{j}}$ and $Y_{\bar{j}} \in \Omega$. Clearly j must be adjacent to i, but we may make a stronger statement than this. Suppose $i = (a, b)$. Then if $\delta(a) = 3$, there are at most 2 such choices of j incident to a, since there are then only 2 edges incident to a that are not i. If $\delta(a) > 3$, however, there is at most one such choice of j. To see this, observe that in both X and Y there must be at least 2 edges other than i either oriented towards or away from a, thus there can only be one edge incident to a that is critical for a.

Thus, if G has minimum degree 4, then

$$|\{j \in V \mid j \neq i; \text{ precisely one of } X_{\bar{j}} \text{ and } Y_{\bar{j}} \in \Omega\}| \leq 2,$$

and thus $\beta \leq 1$ and the Metropolis Path Coupling Theorem applies.

We may improve this result slightly, to cover the class of graphs in which no vertex of degree 3 is adjacent to another vertex of degree 3. Consider the following Markov chain:

1. choose an edge, e, uniformly at random.

2.   • if e is incident to a vertex of degree 3

   with probability $5/6$ do nothing, and with probability $1/6$ choose new orientations for all 3 edges incident to this vertex, uniformly at random from the set of (up to 6) orientations that are permissible for these 3 edges, given their surroundings (i.e. a heat-bath move).

      • otherwise

   choose a new orientation for e, uniformly at random from the set of (up to 2) orientations that are permissible given the orientations of the adjacent edges.

Once again, it is clear that this Markov chain is ergodic with stationary distribution uniform on the set of sink- and source-free orientations of G.

We will analyse this Markov chain by path coupling: once more we will consider only pairs of states that differ in their orientation of exactly one edge, and take the Hamming distance as our metric.

The coupling we use is simple: choose an edge e uniformly at random, and, then choose to couple any subsequent choice of orientations "optimally" [25], i.e. so as to minimize the expected distance between the two coupled copies.

The analysis of this Markov chain is straightforward. If we are considering two states, X and $Y = X_{i \to c}$ for some i and c, there are three cases:

1. The edge i is adjacent to a vertex of degree 3.

    Here, the probability of the distance decreasing to zero is $1/2m$, and the probability of the distance increasing is zero.

2. The edge i is adjacent to a critical edge, and that edge is adjacent to a vertex of degree 3. (There can be at most two such critical edges.)

    Here, the probability of the distance decreasing is $1/m$. The distance could be increased by choosing any one of the (up to 6) edges incident to the vertex or vertices of degree 3 mentioned. For each of these, there is a probability of $1/6$ of actually choosing a new orientation. There are very few cases to consider, and the worst case is where there are only three possible orientations in X and one in Y. In this case there would be a probability $1/3$ of the distance increasing by 1 and $1/3$ of it increasing by 2. Overall therefore, the expected change in distance is zero.

3. All other cases.

    Here the analysis is identical to that of the first Markov chain presented in this section. Overall in this case, the expected change in distance is once again zero.

Applying Proposition 12 and Conclusion 2 of the Metropolis Path Coupling Theorem (with $\eta = 1$), we see immediately that $\mathcal{M}$ is rapidly mixing with mixing rate $2e(n^2 m + m^3 + 2nm^2 + nm + m^2)\lceil \ln \varepsilon^{-1} \rceil = O(m^3 \log \varepsilon^{-1})$.

Since Corollary 13 for sink- and source-free orientations is the exact analogue of Corallary 5 for sink-free orientations, exactly the same reduction from approximate counting holds as for sink-free orientations in Section 4.2.5. Thus there is an *fpras* for sink- and source-free orientations, for graphs of minimum degree 3 in which no two vertices of degree 3 are adjacent, and to graphs of minimum degree 4, which runs in time $O(n^2 m^3 \varepsilon^{-2} \log m \varepsilon^{-1})$.

5.4 Totally Edge Cyclic Orientations

In this section we consider the set of totally edge cyclic orientations of a graph. A totally edge cyclic orientation of a graph is an assignment of a direction to each of the edges of the graph in such a fashion that every edge lies on a directed cycle.

A strong orientation of a graph is an assignment of a direction to each of the edges of the graph such that there is a directed path between every pair of vertices of the graph. The existence of a strong orientation of a graph was resolved by Robbins [104]. Robbins showed that a strong orientation exists if and only if the graph has no *bridge*. (A bridge is an edge, which, if deleted, would disconnect the graph.) Robbins goes on to show how to construct a strong orientation provided one exists.

For connected graphs, an orientation is a strong orientation if and only if it is totally edge cyclic.

Since the number of totally edge cyclic orientations is equal to the Tutte polynomial of the graph evaluated at $(0, 2)$, it follows from Jaeger, Vertigan, and Welsh [65] that evaluating their number is in general #P-complete.

Alon, Frieze, and Welsh [6] gave an *fpras* for the number of totally edge cyclic orientations of dense graphs; they reduced approximate counting to sampling, and showed that a direct sampler would have an expected polynomial running time for the case of dense graphs.

In this section we consider a Markov chain on the set of totally edge cyclic orientations. We show by the method of path coupling that this Markov chain is rapidly mixing. This does not, however, result in an approximate sampler for totally edge cyclic orientations, since the Markov chain is not in general irreducible.

On which class of graphs our Markov chain is irreducible remains an open problem. We prove here that for any class of graphs for which the Markov chain is irreducible, the Markov chain is also rapidly mixing.

5.4.1 Approximate Sampling

For a graph G, let $\mathcal{M} = \mathcal{M}(G)$ be the single-site update Metropolis Markov chain for totally edge cyclic orientations, i.e. with transitions modelled by: pick an edge in G and an orientation for it uniformly at random, and if the resulting orientation is a totally edge cyclic orientation, accept this new orientation as the new state, otherwise reject it, and remain at the original orientation. We shall assume, for simplicity, that G is connected and has no bridge.

We may reinterpret this Markov chain in a way more suitable for application of our theorems. Fix an arbitrary orientation of G, say O. Then we may regard the set of totally edge cyclic orientations as a subset of V^C, where V is the set of edges (note this unusual use of notation) and C is the set $\{\text{Agree}, \text{Disagree}\}$; the interpretation of a member of V^C should be that an edge either "Agrees" or "Disagrees" in orientation with O.

We may now apply the Metropolis Path Coupling Theorem (Theorem 5). If X and Y are two orientations that differ only in their orientation of edge i, write $Y = X_{\bar{i}}$.

Using the notation of Theorem 5, we see that (for $j \neq i$),

$$F_j(X, X_{\bar{i}})$$
$$= \frac{1}{2}\left[|A_{X,j}(\text{Agree}) - A_{X_{\bar{i}},j}(\text{Agree})| + |A_{X,j}(\text{Disagree}) - A_{X_{\bar{i}},j}(\text{Disagree})|\right]$$
$$= \begin{cases} 1 & \text{if precisely one of } X_{\bar{j}}, Y_{\bar{j}} \in \Omega, \\ 0 & \text{otherwise.} \end{cases}$$

Thus, we may write down a formula for β:

$$\beta = 1 - 1/n + \max_{X \in \Omega; i \in V} |\{j \in V \mid j \neq i; \text{precisely one of } X_{\bar{j}} \text{ and } Y_{\bar{j}} \in \Omega\}|/2n,$$

where $Y = X_{\bar{i}}$.

Let us consider how large $|\{j \in V \mid j \neq i; \text{precisely one of } X_{\bar{j}} \text{ and } Y_{\bar{j}} \in \Omega\}|$ can be in the maximum.

Suppose, without loss of generality, that X, Y, and $X_{\bar{j}} \in \Omega$, and $Y_{\bar{j}} \notin \Omega$. For notational convenience, let $i = (a, b)$ and $j = (c, d)$. Without loss of generality we shall assume i is oriented from a to b in $Y_{\bar{j}}$ and j is oriented from c to d (also in $Y_{\bar{j}}$).

Now $Y_{\bar{j}} \notin \Omega$ yet $X_{\bar{j}}$ is. Thus i must lie on a directed cycle in $X_{\bar{j}}$, but not in $Y_{\bar{j}}$. Thus follows there is no directed path from b to a in $Y_{\bar{j}}$.

Consider now the cycle containing i in Y. Since this cycle does not exist in $Y_{\bar{j}}$, it follows that it must include j. Thus there is a directed path from b to d and from c to a in each of X, Y, $X_{\bar{j}}$, and $Y_{\bar{j}}$.

Similarly, there can be no directed path from b to a in $Y_{\bar{j}}$, yet there must be directed paths from d to b and from a to c in each of X, Y, $X_{\bar{j}}$, and $Y_{\bar{j}}$.

We may thus partition the edges of X (say) into 4 sets: $\{i, j\}$, the strongly connected sub-component of $X - \{i, j\}$ that includes a, which we shall write as A; the strongly connected sub-component of $X - \{i, j\}$ that includes b, which we shall call B, and the remaining edges (this set includes a directed path from B to A.) Note that the elements of A and B depend only on i and not on j.

Having established this structure, we are now in a better position to establish how large $|\{j \in V \mid j \neq i; \text{precisely one of } X_{\bar{j}} \text{ and } Y_{\bar{j}} \in \Omega\}|$ can be.

Since $X_{\bar{j}} \in \Omega$, the choice of j is restricted to an edge oriented from B to A in X. But since $Y_{\bar{j}} \notin \Omega$, there can be only one such edge. Thus

$$|\{j \in V \mid j \neq i; \text{precisely one of } X_{\bar{j}} \text{ and } Y_{\bar{j}} \in \Omega\}| \leq 1,$$

and hence $\beta \leq 1 - 1/2n$. Applying the Metropolis Path Coupling Theorem, we see that this chain is rapidly mixing if it is ergodic, and $\log(D(\mathcal{M}))$ is polynomial in the size of the problem. We may deal with this latter question by observing that trivially $D(\mathcal{M}) \leq |\Omega| \leq 2^n$.

We have thus established our claim that if $\mathcal{M}$ is rapidly mixing if it is irreducible.

CONJECTURE 1. *If G is sufficiently dense, then $\mathcal{M}$ is irreducible.*

5.5 Independent Sets: The Conserved Hard-Core Model

In this section we give new results on counting independent sets of fixed sizes in graphs. These results were discussed briefly by Dyer and the author [21]. In the statistical physics community, this problem is known as the *conserved hard-core lattice gas model*.

Consider a graph G on n vertices, where each vertex may be either *occupied* (by a single particle) or *unoccupied*. An instance of a conserved hard-core lattice gas on G with s particles is a configuration in which every vertex adjacent to an occupied vertex is unoccupied. The set of occupied vertices in a hard-core lattice gas is (equivalently) an independent set. We show that provided some condition (to be determined) holds on s, counting the number of these independent sets is #P-complete. We go on to give a fully-polynomial almost uniform sampler for these independent sets, provided again that some condition (to be determined) holds on s. The cases for which we illustrate a fully-polynomial almost uniform sampler are #P-complete.

The condition on s for hardness is a stronger condition than that for approximability.

5.5.1 #P-Completeness of Exact Counting

PROPOSITION 13. *For any fixed $\alpha > 0$, the problem of counting independent sets of fixed size s, where $s \le \alpha n / \Delta$, is #P-complete.*

Proof. Define $I(s)$ to be the number of independent sets of size s in G. Construct graphs G_r by augmenting G with r isolated vertices. We will successively calculate the number of independent sets of each size of G. For each size $s < \alpha n / \Delta$, we shall count these directly.

Consider then the number of independent sets of size $s = \lfloor \alpha(n+r)/\Delta \rfloor$ of G_r. This is equal to

$$\sum_{i=0}^{r} \binom{r}{i} I\left(\left\lfloor \frac{\alpha(n+r)}{\Delta} \right\rfloor - i\right).$$

We can thus calculate $I(k)$ for all $k \le \lfloor \alpha(n+r)/\Delta \rfloor$. By performing these calculations for successively larger (but still polynomially bounded) values, we may calculate $I(k)$ for successively larger values of k. Hence we can calculate the number of independent sets of all sizes in G. This latter problem however is #P-complete [125], completing the proof.   □

5.5.2 Approximate Sampling

Consider the set of functions from particles $P = \{1, 2, \ldots, s\}$ to vertices, V. We call such a function ξ a *placement*. We shall define a Markov chain $\mathcal{M}$ on all placements, to which we shall apply the General Path Coupling Theorem (Theorem 4). A particle i is said to be *safe* in ξ if there is no $j \in P$ $(j \ne i)$ such that either $\xi(i) = \xi(j)$ or

$\xi(i) \sim \xi(j)$. (Recall we write $u \sim v$ if vertices u and v are adjacent.) If all particles
are safe in a placement, then the placement describes an independent set of size s.

Transitions of $\mathcal{M}$ will be as follows. Assume the current state is X. Pick $p \in P$
and $v \in V$ uniformly at random. If p would be safe at v, then accept, and move to this
new placement, otherwise reject and remain at X.

Recalling the notation of Theorem 4, we consider $d_{\mathrm{TV}}(\nu_{X,i}, \nu_{Y,i})$ first. Since we
only choose a different vertex if one of X or Y rejects, and this can happen only if we
choose a vertex occupied by, or adjacent to, another particle we have that

$$d_{\mathrm{TV}}(\nu_{X,i}, \nu_{Y,i}) \leq \sum_{j \neq i} \frac{\delta(X(j)) + 1}{n} \leq \frac{(s-1)(\Delta + 1)}{n}.$$

Now consider $d_{\mathrm{TV}}(\nu_{X,j}, \nu_{Y,j})$. We only choose a different vertex here if we
choose a vertex adjacent (or equal to) to the position of i in either X or Y. Thus
we have

$$d_{\mathrm{TV}}(\nu_{X,j}, \nu_{Y,j}) \leq \frac{\delta(X(i)) + 1 + \delta(Y(i)) + 1}{2n} \leq \frac{\Delta + 1}{n}.$$

Thus

$$\beta \leq 1 - 1/s + 2(\Delta + 1)(s - 1)/sn,$$

and a sufficient condition for $\beta < 1$ is $s < n/2(\Delta + 1) + 1$.

Applying the General Path Coupling Theorem establishes that we have rapid mixing to the uniform stationary distribution on independent sets of size s, provided that
$s < n/2(\Delta + 1) + 1$.

5.6 Independent Sets: The Non-Conserved Hard-Core Model

In this section we give some simple results on counting independent sets of all sizes
in graphs, and a weighted variant of this problem from statistical physics, the (non-
conserved) *hard-core lattice gas model*.

Consider a graph $G = (V, E)$. In statistical physics terms, each vertex is a potential site for a particle; because of the particles' size or to some repulsive properties,
particles are forbidden from occupying adjacent sites. The system is *non-conserved*,
i.e. open to an external environment, so the number of particles observed in the region
being looked at, G, is not constant. Instead, the probability of any configuration with
s particles being observed is proportional to λ^s, where λ is some physically relevant
constant. When $\lambda = 1$, the distribution of states is uniform over all independent sets
in G.

This problem is also considered by Luby and Vigoda [90, 91], Dyer and Greenhill [43], and Randall and Tetali [103]. They analyze different Markov chains from
the one we consider here. Except for a limited class of graphs their results are more
general than those presented here.

This problem is very hard in general — indeed, unless NP $=$ RP, there cannot be a fully-polynomial approximate sampler even in the $\lambda = 1$ case, for general graphs [111]. We provide a fully-polynomial approximate sampler for general graphs, but subject to a strict technical condition on λ.

Exactly counting all independent sets is known to be #P-complete, even if the class of graphs is restricted to those of maximum degree 4 [122, 123]. Indeed, this result has recently been tightened to show #P-completeness even if the class of graphs is restricted to those of maximum degree 3 [43].

5.6.1 Approximate Counting

Given $f : V \to (0, \infty)$, define $f'(v) = \sum_{w:w\sim v} f(w)$. We will define a Markov chain, $\mathcal{M}$, that mixes rapidly for values of λ that satisfy $\lambda[f'(v) - f(v)] \leq f(v)$, for all v. Observe that this holds for *any* positive real-valued function, f.

Let π be the distribution on subsets of V, that is zero except on independent sets, for which it is proportional to λ^s for an independent set of size s.

For convenience, we will assume that each state of the Markov chain is a function from V to the set $\{\text{in}, \text{out}\}$. The interpretation of such a function is as the set of vertices that map to 'in'.

The Markov chain will have transitions defined by the following: suppose the current state is X. Pick $v \in V$ with probability proportional to f, and $c \in \{\text{in}, \text{out}\}$ uniformly at random. Then with probability $\min\{1, \pi(X_{v\to c})/\pi(X)\}$ make a transition to $X_{v\to c}$, and otherwise remain at X.

This Markov chain is clearly a Metropolis Markov chain, and ergodic with stationary distribution π.

We consider two cases: first the case where $\lambda \leq 1$. Applying the Metropolis Path Coupling Theorem, and using notation from its statement, we see that: $S(X, Y) = (1 + \lambda)/2$, $F_j(X, Y) = 0$ for j not adjacent to i, and $F_j(X, Y) \leq \lambda/2$ for j adjacent to i. Thus

$$\beta \leq \max_{v \in V}\{1 - f(v)(1 + \lambda)/2Z + f'(v)\lambda/2Z\},$$

i.e. $\beta \leq 1$ provided $\lambda[f'(v) - f(v)] \leq f(v)$, for all v as required.

In the second case, we consider $\lambda > 1$. Applying the Metropolis Path Coupling Theorem, we have that $S(X, Y) = (1 + \lambda^{-1})/2$, $F_j(X, Y) = 0$ for j not adjacent to i, and $F_j(X, Y) \leq 1/2$ for j adjacent to i. Thus

$$\beta \leq \max_{v \in V}\{1 - f(v)(1 + \lambda^{-1})/2Z + f'(v)/2Z\},$$

i.e. $\beta \leq 1$ provided $\lambda[f'(v) - f(v)] \leq f(v)$, for all v, once more as required.

To illustrate that this Markov chain is rapidly mixing for ranges of λ that cannot be dealt with by either the analyses of Luby and Vigoda [91], or Dyer and Greenhill [43], consider the star graph on $r + 1$ vertices (i.e. the graph $K_{1,r}$).

Choose

$$f(v) = \begin{cases} \frac{1}{\sqrt{r}+1} & \text{if } v \text{ is the centre vertex,} \\ \frac{1}{\sqrt{r}(\sqrt{r}+1)} & \text{otherwise.} \end{cases}$$

Then the result of this section yields rapid-mixing for $\lambda \le 1/(\sqrt{r} - 1)$, whereas that of Luby and Vigoda [91] yields rapid-mixing only for $\lambda \le 1/(r - 3)$; Dyer and Greenhill's [43] result yields rapid-mixing for $\lambda \le 2/(r - 2)$.

5.7 Linear Extensions of a Partial Order

In this section we examine the problem of sampling (almost) uniformly from the set of linear extensions of a partial order, a classic problem in the theory of approximate sampling. Previous techniques have relied on deep geometric arguments, or have not worked in full generality. Recently, focus has centred on the Karzanov and Khachiyan Markov chain. In this section, we define a slightly different Markov chain, and present a very simple proof of its rapid-mixing, using the method of path coupling. We show that this chain has mixing time $O(n^3 \log n)$, which significantly improves the previous best bound for this problem, which was a bound of $O(n^5 \log n)$, for the Karzanov–Khachiyan chain. Other versions of the work in this section have appeared as [19, 22, 23].

5.7.1 Introduction

Let $N = \{1, 2, \ldots, n\}$, and $P = (N, \preceq)$ be a partial order. A *linear extension* of P is a total order $X = (N, \sqsubseteq)$ which respects P, i.e. for all $i, j \in N$, $i \preceq j$ implies $i \sqsubseteq j$. Let $\Omega = \Omega(P)$ denote the set of all linear extensions of P.

Being able to sample from Ω has a variety of applications, since various other combinatorial structures exhibit a natural isomorphism to Ω for a particular family of partial orders, e.g. multiset permutations. In addition there are more direct applications to near-optimal sorting, and to decision theory.

There has been much research on the problem of listing all linear extensions; in fact Pruesse and Ruskey [102] have shown that this may be accomplished in constant amortized time.

Brightwell and Winkler [17] showed that determining $|\Omega|$ is #P-complete. The existence of an *fpras* for $|\Omega|$ followed from the work of Dyer, Frieze, and Kannan [42] on volume approximation. Their method is based on the rapid-mixing of a particular geometric Markov chain. Subsequently, Matthews [92] gave a somewhat different geometric approach.

Karzanov and Khachiyan [80] used geometric and conductance arguments to show the rapid-mixing of a *combinatorial* Markov chain on Ω. Dyer and Frieze [40] improved the conductance estimate, and hence the bound on the mixing time, of this chain. These results all rely on a relationship between Ω and the geometry of a certain polytope in $\mathbb{R}^n$.

Felsner and Wernisch [48] showed how the Karzanov–Khachiyan chain may be used to sample exactly from Ω in the very restricted case of two-dimensional partial orders, using the technique of coupling from the past.

In this section, we significantly reduce the generation time for random linear extensions. We achieve this via the first non-geometric proof of rapid-mixing of a Markov chain on Ω, employing the method of path coupling. We define a new Markov chain on Ω, and show that this has a mixing rate of $O(n^3 \log n\varepsilon^{-1})$, which significantly improves the best bound previously known for this problem (a bound on the Karzanov–Khachiyan chain of $O(n^4 \log |\Omega|\varepsilon^{-1}) = O(n^5 \log n + n^4 \log \varepsilon^{-1})$ [40]). As a consequence, we note that the mixing rate for the Karzanov–Khachiyan chain can in fact be reduced to $O(n^4 \log^2 n + n^3 \log n \log \varepsilon^{-1})$. Wilson [128] has recently shown how the proof in this section may be extended to show that the Karzanov–Khachiyan chain has in fact a mixing rate of $\Theta(n^3 \log n \log \varepsilon^{-1})$.

5.7.2 Notation and Preliminaries

Let $\sigma(i,j)$ $(1 \leq i < j \leq n)$ denote the *transposition* operators on total orders of N. Thus, if $Y = \sigma(i,j)X$, we have

$$X = (a_1\, a_2\, \cdots\, a_{i-1}\, a_i\, \cdots\, a_j\, a_{j+1}\, \cdots\, a_n),$$
$$Y = (a_1\, a_2\, \cdots\, a_{i-1}\, a_j\, \cdots\, a_i\, a_{j+1}\, \cdots\, a_n).$$

If $j = i+1$, we will call $\sigma(i,j)$ a *close* transposition. The integer $(j-i)$ will be called the *width* of the transposition $\sigma(i,j)$. Thus close transpositions have unit width.

For a concave probability distribution, f, on $\{1, 2, \ldots, n-1\}$, define the Markov chain $\mathcal{M}_f$ on $\Omega(P)$: if the current state is $X_t \in \Omega$, then the next state, X_{t+1}, is determined by the following experiment:

1. Choose $p \in \{1, 2, \ldots, n-1\}$ according to the distribution f, and $c \in \{0, 1\}$ uniformly at random.

2. If $c = 0$ or $\sigma(p, p+1)X_t \notin \Omega$, then $X_{t+1} = X_t$. Otherwise $X_{t+1} = \sigma(p, p+1)X_t$.

The symmetry of the transitions implies that the equilibrium distribution, π, is the uniform distribution. We show below, by using the method of path coupling, that $\mathcal{M}_f$ is rapidly-mixing, i.e. μ_t, the distribution of X_t, "quickly" approaches π.

When f is the uniform distribution, we shall abbreviate $\mathcal{M}_f$ to $\mathcal{M}$. Observe that $\mathcal{M}$ is simply the original Karzanov–Khachiyan Markov chain.

For notational convenience we shall assume that f is also defined on 0 and n, and that $f(0) = f(n) = 0$.

Let a *transposition sequence* from X to Y, for any $X, Y \in \Omega$, be any sequence $Z_0, Z_1, \ldots, Z_r$, where $X = Z_0$ and $Y = Z_r$, such that for $k = 1, 2, \ldots, r$, $Z_k \in \Omega$ and $Z_k = \sigma(i_k, j_k)Z_{k-1}$. The integer r is the length of the sequence, and its *weight* is its total width, $\sum_{k=1}^{r}(j_k - i_k)$. The *transposition distance* $\delta(X, Y)$ is then the least weight of any transposition sequence from X to Y. Clearly δ is a metric on Ω. The diameter $D = \max_{X,Y \in \Omega} \delta(X, Y)$ is bounded above by the number of incomparable pairs in P, since there is a sequence of close transpositions of at most this length, and thus $D \leq \binom{n}{2}$.

It transpires that we may in fact find a tighter upper bound on D than this, since it turns out that the transposition distance metric is equivalent to a classical metric on total orders: Spearman's footrule [114]. Suppose X is a total order; we shall use the notation $X(i)$ to denote the the ith element of the total order, X. Then Spearman's footrule, $\delta_S(X, Y) = \frac{1}{2}\sum_{i=1}^{n} |X(i) - Y(i)|$. We prove in Appendix C that $\delta_S = \delta$. It is well known that $\delta_S(X, Y) \leq \lfloor n^2/4 \rfloor$ (see, for example, Diaconis and Graham [34]), and thus $D \leq \lfloor n^2/4 \rfloor$.

5.7.3 The Coupling

Let X and Y be two copies of $\mathcal{M}_f$. At time t, let $X_t = Z_0, Z_1, \ldots, Z_r = Y_t$ be a transposition sequence of minimal weight, d_t. We let the Z_k evolve for a single time step as coupled copies of $\mathcal{M}_f$. Let Z'_k be the state to which Z_k evolves. We couple the Z_k as follows:

1. Choose $p \in \{1, 2, \ldots, n-1\}$ according to the distribution f, and $c_0 \in \{0, 1\}$ uniformly at random.

2. For each $k \in \{1, 2, \ldots, r\}$: if $j_k - i_k = 1$ and $p = i_k$, then let $c_k = 1 - c_{k-1}$. Otherwise let $c_k = c_{k-1}$.

3. For each $k \in \{0, 1, \ldots, r\}$: if $c_k = 0$ or $\sigma(p, p+1)Z_k \notin \Omega$, $Z'_k = Z_k$. Otherwise $Z'_k = \sigma(p, p+1)Z_k$.

We will show that $\mathbf{E}(d_{t+1}) < d_t$, for a suitable choice of f (and in fact that $\mathbf{E}(d_{t+1}) \leq d_t$ for an arbitrary concave f). For notational simplicity, let us write $A = Z_{k-1}$, $B = Z_k$ and $(i, j) = (i_k, j_k)$. Thus $B = \sigma(i, j)A$, i.e.

$$A = (a_1\, a_2\, \cdots\, a_{i-1}\, a_i\, \cdots\, a_j\, a_{j+1}\, \cdots\, a_n),$$
$$B = (a_1\, a_2\, \cdots\, a_{i-1}\, a_j\, \cdots\, a_i\, a_{j+1}\, \cdots\, a_n).$$

If $p \notin \{i-1, i, j-1, j\}$ then $\delta(A', B') = (j - i) = \delta(A, B)$, since either we do nothing in both A and B, or $\sigma(p, p+1)$ can be successfully applied in both A and B.

If $p = i - 1$, then either we do nothing in both A and B with probability $\frac{1}{2}$, or we attempt to make the transposition, $\sigma(i-1, i)$, in both A and B. If this transposition is successful in neither, we will have $\delta(A', B') = \delta(A, B)$; if it is successful in both, then we will have $\delta(A', B') = \delta(A, B) + 1$, since A' and B' will differ by a transposition of width $(j - i + 1)$. If it is successful in only one of A and B then $\delta(A', B') = \delta_S(A', B') = \delta(A, B) + 1$. Thus, conditional on $p = i - 1$, or similarly $p = j$, we have $\mathbf{E}(\delta(A', B')) \leq \delta(A, B) + \frac{1}{2}$.

Now consider $p = i$. Suppose first that $i \neq j - 1$, i.e. $(j - i) > 1$. Then the transposition, $\sigma(i, i+1)$, must succeed in both A and B — for suppose to the contrary that it fails, in A, say. Then $a_i \preceq a_{i+1}$, but a_i follows a_{i+1} in B, so $B \notin \Omega$, establishing a contradiction. Thus with probability $\frac{1}{2}$ we have $\delta(A', B') = \delta(A, B) - 1$, since A', B' will differ by a transposition of width $(j - i - 1)$. Thus conditional on $p = i$, or similarly, $p = j - 1$, we have $\mathbf{E}(\delta(A', B')) \leq \delta(A, B) - \frac{1}{2}$. Therefore, if $i \neq j - 1$,

$$\mathbf{E}(\delta(A', B')) - \delta(A, B) \leq \frac{1}{2}(f(i-1) - f(i) - f(j-1) + f(j)).$$

It remains only to consider the case $p = i = j - 1$. Clearly we can apply $\sigma(i, i+1)$ in both A and B, since $B = \sigma(i, i+1)A$. Moreover, the coupling $c(B) = 1 - c(A)$ ensures that we do nothing in one and transpose in the other. Thus $B' = A'$ with unit probability, and $\delta(A', B') = \delta(A, B) - 1$. Therefore, if $i = j - 1$,

$$\mathbf{E}\left(\delta\left(A', B'\right)\right) - \delta(A, B) \leq \tfrac{1}{2}\left(f(i-1) + f(j)\right) - f(i)$$
$$= \tfrac{1}{2}\left(f(i-1) - f(i) - f(j-1) + f(j)\right).$$

So, in all cases, we have that the unconditioned expectation, $\mathbf{E}\left(\delta(A', B')\right) \leq \delta(A, B) + \tfrac{1}{2}(f(i-1) - f(i) - f(j-1) + f(j))$.

Note that this shows that the transposition distance does not increase in expectation under this coupling for *any* concave probability distribution f. It is possible to prove rapid-mixing in this general setting, resulting in a mixing time of $O(n^5)$, however we do not do this here. Instead we fix on a particular choice of f and show that, for this f, $\mathcal{M}_f$ has a mixing time of $O(n^3 \log n)$.

Our choice of concave probability distribution is $F(i) = i(n-i)/K$, where K is the normalizing constant. It is easy to verify that $K = \tfrac{1}{6}\left(n^3 - n\right)$. We choose F as quadratic, since we observe that to minimize $\max_{i<j}\{(f(i-1) - f(i) - f(j-1) + f(j))/(j-i)\}$, we should use a function with a constant second difference.

Then, for all i, $F(i) - F(i-1) = (n+1-2i)/K$, and thus $\tfrac{1}{2}(F(i-1) - F(i) - F(j-1) + F(j)) = (i-j)/K$.

Now recall that $\delta(A, B) = j - i$, and thus $\mathbf{E}\left(\delta(A', B')\right) \leq (1 - \tfrac{1}{K})\delta(A, B)$.

Applying the Path Coupling Lemma we see that $d_t \leq \left(1 - \tfrac{1}{K}\right)^t D$, and since d_t is a non-negative integer-valued variable, $\mathbf{P}(X_t \neq Y_t) \leq \left(1 - \tfrac{1}{K}\right)^t D$. Applying the Coupling Lemma, we see that $d_{TV}(X_t \neq Y_t) \leq \left(1 - \tfrac{1}{K}\right)^t D$. Taking logarithms, and rearranging, we see that in order to ensure that $d_{TV}(X_t \neq Y_t) \leq \varepsilon$ we need only simulate $\mathcal{M}_F$ for $K \ln D\varepsilon^{-1}$ steps.

Recalling that $D \leq \lfloor n^2/4 \rfloor$, we see that this shows that the mixing rate of $\mathcal{M}_F$ is bounded above by $\lceil \tfrac{1}{6}(n^3 - n) \ln(n^2 \varepsilon^{-1}/4) \rceil$. As $d_{TV}(X_t \neq Y_t) \leq \left(1 - \tfrac{1}{K}\right)^t D$, it follows from Lemma 1 that the relaxation time of $\mathcal{M}_F$ is $\tfrac{1}{6}(n^3 - n)$.

5.7.4 Lower Bounds and Related Chains

Wilson [128] has shown a lower bound on the mixing times of the Karzanov–Khachiyan chain of $\Omega(n^3 \log n)$; the same methods may be applied [127] to provide an $\Omega(n^3 \log n)$ lower bound on the mixing time of $\mathcal{M}_F$ too. The results on $\mathcal{M}$ and $\mathcal{M}_F$, are summarized in Table 5.2.

Whether the lower bound is $\Omega(n^3 \log n)$ for an arbitrary convex f remains open, however it is easy to show a lower bound of $\Omega(n^3)$ in this generality: consider the partial order consisting of a chain of length $n - 1$, and one independent element. This would have only n linear extensions: the chain, with the independent element inserted at each point. $\mathcal{M}_f$ would thus in this case be equivalent to a random walk on $\{1, 2, \ldots, n\}$, denoting the position of the random element, and moving with probabilities given by f. Since f is a concave probability distribution, its maximum is at most $2/n$. To see this, suppose the maximum is $f(k) = h$. Then $f(i) \geq hi/k$ for

Table 5.2: A summary of the mixing and relaxation times.

<table>
<tr><td colspan="2" align="center">$\mathcal{M}_F$</td></tr>
<tr><td align="center">Lower bounds</td><td align="center">Upper bounds</td></tr>
<tr>
<td align="center">$\tau = \Omega(n^3 \log n)$

(From [127])

$\tau_2 \geq \dfrac{1}{\pi^2} n^3 \approx 0.101 n^3$

(Follows from [35] with [43])</td>
<td align="center">$\tau \leq \left(\dfrac{\ln 2}{6} + o(1) \right) n^3 \ln n$

$\approx 0.116 n^3 \ln n$

$\tau_2 \leq {}^{1}\!/\!_{6}(n^3 - n)$

$\approx 0.167 n^3$

(This section)</td>
</tr>
</table>

<table>
<tr><td colspan="2" align="center">$\mathcal{M}$, the Karzanov–Khachiyan chain.</td></tr>
<tr><td align="center">Lower bounds</td><td align="center">Upper bounds</td></tr>
<tr>
<td align="center">$(1/\pi^2 - o(1))n^3 \ln n \leq \tau$

$\approx 0.101 n^3 \ln n$

(Wilson [128])

$\tau_2 \geq \dfrac{2}{\pi^2} n^3 \approx 0.203 n^3$

(Follows from [35])</td>
<td align="center">$\tau \leq \left(\dfrac{4}{\pi^2} + o(1) \right) n^3 \log n$

$\approx 0.405 n^3 \log n$

(Wilson [128], following [22])

$\tau_2 \leq {}^{1}\!/\!_{3}(n^3 - n)$

$\approx 0.333 n^3$

(This section, with [43])</td>
</tr>
</table>

$i \leq k$ and $f(i) \geq h(n-i)/(n-k)$ for $i > k$. Thus

$$1 = \sum_{i=1}^{n-1} f(i) \geq \sum_{i=1}^{k} hi/k + \sum_{i=k+1}^{n-1} h(n-i)/(n-k) = hn/2,$$

from which the assertion follows. Thus the expected time before $\mathcal{M}_f$ could perform a non-null transition is $\Omega(n)$. Since a simple random walk on $\{1, 2, \ldots, n\}$ has mixing time $\Theta(n^2)$, we see that for any concave choice of f, $\mathcal{M}_f$ has a mixing time that is $\Omega(n^3)$ in the worst case.

There is therefore no substantial complexity gap between the mixing time proved in this section and the theoretical optimum for this class of chains. Any significant improvement over the results of this section would have to be made by turning to different Markov chains.

A corollary of the mixing rate proved in this section, together with some work of Dyer and Greenhill [43] on eigenvalue comparisons, is that the mixing rate of the Karzanov–Khachiyan chain is worse than that of $\mathcal{M}_F$ by no more than a factor of $O(\log |\Omega|) = O(n \log n)$; i.e. that the mixing rate of $\mathcal{M}$ is $O(n^3 \log n \log |\Omega| \varepsilon^{-1})$ (an improvement on the bound previously known of $O(n^4 \log |\Omega| \varepsilon^{-1})$ [40]), and furthermore, that the Karzanov–Khachiyan chain has a relaxation time that is worse than that of $\mathcal{M}_F$ by no more than a constant factor — both are $\Theta(n^3)$ and therefore optimal on this criterion. Subsequent to an earlier version of the results of this section, Wilson [128] showed how the results of a version of this section [22] may be extended to the Karzanov–Khachiyan chain directly, to yield a mixing time of $\Theta(n^3 \log n)$.

An obvious candidate for an improved chain would be that which performs transitions by choosing a random transposition rather that merely a random close transposition. On the set of all total orders this is known to have a mixing time of only $\Theta(n \log n)$ [37]. Currently, the best known bound on the mixing time of this chain in general is $O(n^4 \log n \log |\Omega| \varepsilon^{-1})$, and the relaxation time is $O(n^4)$. (This fact is derived from the mixing rate proved in this section, again together with some work of Dyer and Greenhill [43]). This approach cannot hope to improve the result in this section, however, for the relaxation time (and hence the mixing time) of this chain is *in general* $\Omega(n^4)$ (consider again the partial order that consists of a chain and an element; here the probability of a transition being non-null is $O(1/n^2)$).

A more promising candidate is the chain that performs transitions by choosing a random element to delete and a random position for its subsequent reinsertion. This has a mixing time of $\Theta(n \log n)$ on the set of all total orders. Again, the mixing rate proved in this section together with [43] shows an upper bound on the mixing rate of this chain of $O(n^4 \log n \log |\Omega| \varepsilon^{-1})$, and relaxation time of $O(n^4)$. In general, we only know how to show a lower bound of $\Omega(n^2)$ for the mixing time of this chain. Consider the partial order that is formed from two independent chains, A and B, each of size $(2n + 1)$. Let a and b respectively be the middle elements of the two chains. If we were to start this Markov chain with the ordering in which all elements of A are ranked before all elements of B, then the initial distance between a and b is $(n + 1)$. Consider how this distance may decrease. In order for it to decrease we would first have to select either the least element of A, or the largest element of B, between a and b. The expected time before this occurs is $\Theta(n)$. Since the initial distance between a

and b is $\Theta(n)$, the expected time before we may have b before a is $\Omega(n^2)$ — but since there are as many linear extensions with b before a as b after a, we see that $\Omega(n^2)$ is a lower bound on the mixing time.

One of the major applications of the generation of random linear extensions is to approximate the *number* of linear extensions. Calculating this number exactly is #P-complete [17]. Brightwell and Winkler [17] illustrate an *fpras* for approximating this number, i.e. an algorithm that approximates the exact number to within a multiplicative factor of $1 + \varepsilon$ with probability at least $3/4$: this uses an almost uniform sampler as a subroutine; the running time of this algorithm, using the original bounds on the Karzanov–Khachiyan chain, is $O(n^9 \log^6 n \, \varepsilon^{-2} \log \varepsilon^{-1})$. Dyer and Frieze [40], as well as improving the bound on the mixing rate of the Karzanov–Khachiyan chain, use an improved algorithm for the approximate counting problem that runs in time $O(n^6 \log^2 n \, \varepsilon^{-2} \log(n\epsilon^{-1}))$. Using what are now standard techniques (see e.g. [3, 70]), if we have a Markov chain with stationary distribution the uniform distribution on the set of linear extensions, we may use this to generate an algorithm for approximating the number of linear extensions that will run in time $O(n^2 \log^2 n \, \varepsilon^{-2} \log(n\varepsilon^{-1}) \times \tau_2 + \tau(\varepsilon/n \log n))$. In particular, for $\mathcal{M}$, or $\mathcal{M}_\mathcal{F}$, this translates to an approximation scheme that runs in time $O(n^5 \log^2 n \, \varepsilon^{-2} \log(n\varepsilon^{-1}))$.

5.7.5 Conclusions

We have shown a significantly improved bound on the mixing rate of Markov chains for generating random linear extensions of a partial order. We have also indicated that our results are close to optimal for a family of chains which includes the Karzanov–Khachiyan chain as a special case. The convergence proof is non-geometric, in sharp contrast to all earlier approaches to this problem, and serves well to illustrate the utility of the path coupling method in an arena that has been extensively analysed by other methods.

We round off this section with a few conjectures.

CONJECTURE 2. *The entire family of Markov chains, $\mathcal{M}_f$, for convex f has mixing time $\Theta(n^3 \log n)$, and relaxation time $\Theta(n^3)$.*

CONJECTURE 3. *The worst-case mixing time for the Karzanov–Kachiyan Markov chain is achieved when Ω is the set of linear extensions of the anti-chain (i.e. Ω is the set of all permutations).*

5.8 Graph Colouring

In this section, we consider the problem of sampling (almost) uniformly from the set of proper k-colourings of a graph. This problem has been considered by a variety of authors, with the original proofs of the existence of a fully-polynomial almost uniform sampler coming independently from Jerrum [67] and Salas and Sokal [106]. They show the existence of such a sampler for $k \geq 2\Delta + 1$, where Δ, as usual, is the maximum degree of the graph. In Section 4.1, we generalized these results both to $k \geq 2\Delta$, and from graphs to hypergraphs. In this section, we show how the result of $k \geq 2\Delta$ may be regained with far more ease using the technique of path coupling. We consider an extension to hypergraphs and the anti-ferromagnetic Potts model in Section 5.9.

For a graph $G = (V, E)$ and set (of colours) C, a function $\chi : V \to C$ is said to be a *colouring*. We will say a vertex v is properly coloured if $\chi(v) \notin \{\chi(w) : w \sim v\}$. A *proper colouring* of G is a colouring with all vertices properly coloured.

Jerrum [67] exhibits a fully-polynomial almost uniform sampler for k-colourings of a graph, provided that $k \geq 2\Delta + 1$.

Consider the Markov chain $\mathcal{M}$ with state space the set of all colourings of G and transitions, at state X, defined as follows:

1. Choose v uniformly at random from V, and c uniformly at random from C.

2. If v is properly coloured in $X_{v \to c}$, then $X' = X_{v \to c}$, otherwise $X' = X$.

This is an extension of Jerrum's chain to all of C^V. It is easy to show that the positive-recurrent states of $\mathcal{M}$ are the proper colourings of G, and that the chain is ergodic on these states, but we omit these details here. Let us apply Theorem 4. In this instance, J is the constant function with value $1/n$, and unless $i \sim j$ or $i = j$, $\nu_{X,j} = \nu_{Y,j}$. In the case $i = j$, we have $d_{\mathrm{TV}}(\nu_{X,j}, \nu_{Y,j}) \leq \Delta/k$, since any colour choice that would be accepted in X would also be accepted in Y. Furthermore, for $j \sim i$, $d_{\mathrm{TV}}(\nu_{X,j}, \nu_{Y,j}) \leq 1/k$, since every colour that would be accepted in X (or Y), except possibly $Y(i)$ (or $X(i)$, respectively), would be accepted in Y (or X, respectively). Thus $\beta \leq 1 - (1 - \Delta/k)/n + \sum_{j \sim i} 1/kn \leq 1 - 1/n + 2\Delta/kn$.

Thus, applying Theorem 4, we see that $\mathcal{M}$ is rapidly mixing for $k \geq 2\Delta + 1$, and thus comes within ε of its stationary distribution after at most $\lceil \ln(n\varepsilon^{-1}) / \ln((kn - k + 2\Delta)/kn) \rceil$ steps.

However, we can possibly do better than this. We need not take J to be constant, and in general, $\beta \leq \max_i \{1 - J(i)(1 - \delta(i)/k) + \sum_{j \sim i} J(j)/k\}$. So if, for example, we take $J(i) = \delta(i)/2m$, where $m = |E|$, we see that the chain would be rapidly mixing provided $k > \max_{v \in V} \{\delta(v) + \sum_{w \sim v} \delta(w)/\delta(v)\}$, i.e. the largest degree of a vertex plus the average degree of its neighbours. This will be less than 2Δ unless there is a vertex of maximum degree for which all of its neighbours are also of maximum degree. Thus, in particular, for a bipartite graph with maximum degrees Δ_L and Δ_R respectively for its two bipartitions, a sufficient condition for rapid mixing is to take $k > \Delta_L + \Delta_R$.

It is natural to inquire whether we can choose J optimally for a particular graph, and it is not hard to see that we may in fact do so. Answering the query, "is there a J such that $\mathcal{M}$ is provably rapidly mixing for k colours using this proof technique?" is equivalent to checking that a particular linear program is feasible, and finding such a J is equivalent to finding a feasible solution. Specifically, J must satisfy the following constraints:

$$\sum_{i \in V} J(i) = 1,$$

$$1 - J(i)(1 - \delta(i)/k) + \sum_{j \sim i} J(j)/k \leq 1 \quad \text{for all } i \in V.$$

It is also possible to show convergence in the case $k = 2\Delta$, as noted in [67] using the remark following Theorem 4. The only stuck pairs are certain proper colourings X, Y with $H(X, Y) = n$. If we simply modify the coupling to allow them to evolve independently, they cannot remain stuck for long. We omit the details here (which were covered in Section 4.1.2.4), and simply note that in this case the mixing rate is $O(kn^3 \log \varepsilon^{-1})$.

5.9 The Extended Potts Framework

The following subsumes and generalizes Section 5.8, but with a slightly more complicated approach.

Many systems in statistical physics are referred to as *anti-ferromagnetic systems*, where more energy is required for a system in which adjacent particles have the same (or in some cases similar) states.

The *Hamiltonian* $\mathcal{H}$ defines the *energy* of a state. According to the axioms of statistical mechanics, if an anti-ferromagnetic system is in equilibrium with surroundings at a temperature T, then the probability of observing any particular state χ is proportional to $\exp(-\gamma\mathcal{H}(\chi))$, where[3] $\gamma = 1/kT$, and k is a constant. The probability distribution governing the observed states is known as the *Gibbs distribution*.

The *k-state Potts model* for anti-ferromagnetism assigns one of k "spins" (colours) to each of the vertices in a graph. The Hamiltonian, $\mathcal{H}$, is defined to be equal to the number of flaws in the colouring.

Recall that a hypergraph is a set of vertices, together with a set of "edges"—each edge can contain any number of vertices, but to avoid trivialities, we shall assume here that they each contain at least two. (Clearly, if they all contain exactly two we just have a graph.)

As for graphs, we define a colouring to be a function $\chi : V \to C$. The number of *flaws* $f(\chi)$ of χ is the number of edges, $e \in E$, for which all of its vertices are the same colour. A *proper colouring* is a colouring with no flaws. We say colour c is *critical* for $v \in V$ if there is any edge where v is the only vertex not coloured c. We call such an edge a *critical edge*. A hypergraph is k-colourable if it can be properly coloured with k colours.

It should be clear that we may extend the definition of the Potts model to hypergraphs. In this case we will refer to the model as the *extended Potts framework*. It

---

[3] The usual symbol is β, but we use γ here to avoid confusion.

should be noted that the extended Potts framework is not a physical model *per se*, but rather a *framework* within which physical models, such as the Potts model, may be set. Another model which lies within the framework might define a flaw to be a particle which has the same spin as *all* its neighbours.

Consider the Metropolis Markov chain $\mathcal{M}$, defined as follows.

1. At colouring X choose a vertex v and colour c uniformly at random.

2. Accept $X_{v \to c}$ with probability $\min\{1, \exp(f(X) - f(X_{v \to c}))\}$.

It is easily checked that the stationary distribution is the Gibbs distribution for the extended Potts framework.

With a chain similar to $\mathcal{M}$, Salas and Sokal [106] use Dobrushin's Uniqueness Criterion (see also [27], Section 2.3.2, and Appendix A) to show absence of phase transition for the Potts model (which implies rapid-mixing) provided $k > 2\Delta$. Jerrum [67] observed, without proof, that it would be sufficient to take $k > 2(1 - e^{-\gamma})\Delta$.

We apply Theorem 4 with J uniform. Consider first the case where $i = j$. Assume without loss of generality that $\pi(X) \geq \pi(Y)$, and define $\varrho = \mathrm{d_{TV}}(\nu_{X,j}, \nu_{Y,j})$ to be the probability that X rejects. There can be at most $\delta_1(i)$ colours that have a non-zero probability of rejecting, since at most $\delta_1(i)$ colours can be critical for i. Enumerate these by $c_1, c_2, \ldots, c_r$. Then ϱ is bounded above by $\sum_{l=1}^{r}(1 - e^{-\gamma \sigma_l})/k$, where σ_l is the number of edges critical for i with colour c_l. Now, since $1 - e^{-x}$ is a convex increasing function, and $\sum_{l=1}^{r} \sigma_l \leq \delta_2(j)$, we have $\varrho \leq \delta_1(i)[1 - e^{-\gamma \delta_2(i)/\delta_1(i)}]/k$.

For the cases in which $j \neq i$, we have $\mathrm{d_{TV}}(\nu_{X,j}, \nu_{Y,j}) \leq (1 - e^{-\gamma \delta(i,j)})/k$, since there could be at most $\delta(i, j)$ additional flaws for any colour choice at j caused by changing only the colour of i. Also there are at most $\delta_4(i)$ vertices for which $\mathrm{d_{TV}}(\nu_{X,j}, \nu_{Y,j}) > 0$. Any two such vertices must be adjacent to i, but not both in any edge containing i. This follows from the fact that an edge may be critical for at most one vertex other than i in either X or Y.

Thus $\mathcal{M}$ is rapidly mixing provided that

$$\max_{i,j \in V} \left\{ 1 - \frac{1}{n} + \delta_1(i) \frac{1 - e^{-\gamma \delta_2(i)/\delta_1(i)}}{nk} + \delta_4(i) \frac{1 - e^{-\gamma \delta(i,j)}}{nk} \right\} < 1.$$

Thus, in particular, $k > \Delta_1 + \Delta_4$ is a sufficient condition. Also it is possible to show that the only stuck pairs are proper colourings of a *regular graph* of degree Δ, and hence we could show convergence also in the case $k \geq \Delta_1 + \Delta_4$.

In the case of the Potts model, we have $\delta_r = \delta$ $(r = 1, 2, 3, 4)$, and $\delta(i, j) \leq 1$. Thus a sufficient condition is $k > 2\Delta(1 - e^{-\gamma})$. Jerrum [67] mentions that a similar result, for a slightly different Markov chain, has been obtained by Salas and Sokal. Note also that once again the equality case can be handled using the remark following Theorem 4.

5.10 Graph Colouring Revisited

Both Jerrum [67] and Salas and Sokal [106] independently proved that a simple random walk on the colourings of a graph would mix rapidly, provided the number of colours, k, exceeded the maximum degree, Δ, of the graph by a factor of more than 2. These proofs were based on entirely different techniques, *coupling* and *Dobrushin uniqueness* respectively. The two results had different merits. Jerrum's result was subsequently extended (but with an $\Omega^*(n^2)$ increase in running time[4]) to the case $k = 2\Delta$, whereas Salas and Sokal's was extended to the *Potts model* [21]. However the similarity of these bounds suggested a natural conjecture, that 2Δ might be a barrier for rapid mixing of the underlying Markov chain on colourings. (Recall that the chain is known to converge eventually for $k \geq \Delta + 2$.) Indeed (his q being our k) Sokal [113] wrote

> ...it might be ... that $q > 2\Delta$ is in fact the *best possible* result for rapid mixing for arbitrary graphs in terms of only Δ.

(The extension of Jerrum's result naturally weakened the "2Δ conjecture" to $k \geq 2\Delta$.)

We saw in Section 5.9 how a path coupling proof could be extended to the Potts model (and beyond). Dyer and Greenhill [46] have also used path coupling to analyze a more rapidly mixing chain on colourings. This reduces the running time for the case $k = 2\Delta$ by an $\Omega^*(n^2)$ factor, but still does not beat the 2Δ barrier.

Häggström and Nelander [62] have used coupling from the past to provide experimental evidence that on the planar Cartesian lattice the $k = 2\Delta(= 8)$ case is qualitatively much more slowly mixing than the $k > 2\Delta$ case. This certainly seems to support the dichotomy in the bounds in the mixing times that results from applying path coupling in a straightforward fashion (see Section 5.8 — the dichotomy reflects the two cases of $\beta < 1$ and $\beta \leq 1$ in the Metropolis Path Coupling Theorem). However the methods of this section show rapid-mixing for an elementary Markov chain on 7-colourings of triangle-free 4-regular graphs, which implies, via a comparison lemma of Dyer and Greenhill [43], that the chain studied by Häggström and Nelander [62] is also rapidly mixing for $k = 7$.

In this section, we explore a restricted colouring problem in an attempt to break this 2Δ bound. We consider the problem of 5-colouring graphs of maximum degree 3. We show that this problem remains #P-complete, and go on to give a synopsis of a proof of rapid-mixing for a Markov chain on this state space. A full version of this proof (and parts of this introduction) appeared in Bubley, Dyer, Greenhill and Jerrum [25].

5.10.1 #P-Completeness

We show that counting the number of 5-colourings of 3-regular graphs is a #P-complete problem. The proof is based on an idea of Mark Jerrum. A stronger result, showing that k-colouring graphs of maximum degree Δ is #P-complete for $k \geq 3$, $\Delta \geq 3$ appears in [25].

We proceed by a three-step reduction. We reduce counting 5-colourings of an arbitrary graph to a special version of the 5-colouring called Π (which we explain below), and thence to 5-colouring graphs of maximum degree 3, before finally coming to 5-colouring 3-regular graphs.

---

[4]$\Omega^*(\cdot)$ is the notation which hides factors of $\log n$.

Define the problem Π:

> **Instance:** A graph, $G = (V, E)$, of maximum degree 3, with a bipartition of $B \cup M$ of E, the edge set.
>
> **Question:** How many 5-colourings are there of the vertices of G that make all edges in B bichromatic, and all edges in M monochromatic?

LEMMA 27. *5-colouring an arbitrary graph reduces parsimoniously to* Π, *and hence* Π *is a #P-complete problem.*

Proof. Given an instance of Π, contract all edges in M. Since the vertices at the two ends of M are, by definition, to be coloured the same, a Π-colouring of the original graph is a proper 5-colouring of the mapped graph, and vice versa. Furthermore, it is clear that there is an instance of Π that will become, under this mapping, an arbitrary graph. Thus 5-colouring an arbitrary graph reduces parsimoniously to Π.

Note that 5-colouring an arbitrary graph is a #P-complete problem, since it is clearly within #P, and its calculation is essentially the evaluation of a non-special point of the Tutte polynomial (in fact, point $(-4, 0)$), and hence #P-hard [65]. $\square$

We now reduce Π to the problem of 5-colouring graphs of maximum degree 3.

LEMMA 28. *Counting the number of 5-colourings of graphs of maximum degree 3 is #P-complete.*

Proof. Given an instance $G = (V, M \cup B)$, of Π, construct graphs G_k for $k = 1, 2, \ldots$, by substituting a path of length k for each edge in M. So, for example, G_1 is just G without the bipartition of the edge set.

Consider the set of 5-colourings of G, such that all edges of B are bi-coloured, and those of M may be coloured arbitrarily. Let N_s be the number of these colourings that bi-colour s edges of M, and mono-colour the remainder, $m - s$ (where $m = |M|$).

Let d_k be the number of 5-colourings of a path of length k, where the end-points have fixed and different colours. Let e_k be the number of 5-colourings of a path of length k, where the end-points have identical fixed colours.

We may now write the number of proper 5-colourings of G_k as a simple sum:

$$\sum_{s=0}^{m} N_s d_k^s e_k^{m-s} = e_k^m \sum_{s=0}^{m} N_s \left(\frac{d_k}{e_k}\right)^s$$

One may evaluate d_k and e_k using simple combinatorics, e.g. it is easy to see that they must satisfy the following difference equations:

1. $d_k = e_{k-1} + 3d_{k-1}$,

2. $e_k = 4d_{k-1}$,

3. $d_1 = 1$ and $e_1 = 0$.

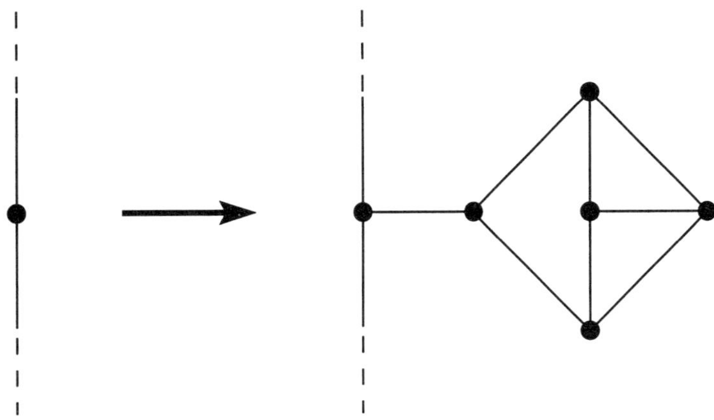

Figure 5.2: Modifying the graph.

Solving these, we have $d_k = (4^k + (-1)^{k+1})/5$, and $e_k = (4^k + 4(-1)^k)/5$. We may easily pick $m + 1$ values of k so that we get $m + 1$ distinct values of d_k/e_k. Thus we can obtain a set of $m + 1$ linear equations in N_s for $s = 0, 1, \ldots, m$. Since the coefficients of these linear equations form a Vandermonde matrix with distinct columns, these equations are linearly independent, and we may thus calculate the values of N_s for $s = 0, 1, \ldots, m$. Note that N_0 is precisely the number of Π-colourings of G. Thus if we could calculate the number of 5-colourings of graphs of maximum degree 3 in polynomial time, we could count solutions of Π in polynomial time. $\quad\square$

Finally, we show that if one could count the number of 5-colourings of 3-regular graphs, one could count the number of 5-colourings of graphs of maximum degree 3.

LEMMA 29. *Counting the number of 5-colourings of 3-regular graphs is #P-complete.*

Proof. Given a graph of maximum degree 3, we may modify this graph to make it regular in such a fashion that the relationship between the number of 5-colourings of the original graph and of the modified graph is known.

We will deal with vertices of degree 1 first: these may simply be deleted, together with the edge upon which they are incident. Each such deletion multiplies the number of proper 5-colourings by a factor of $1/4$.

Secondly, vertices of degree 2 may be replaced by a small subgraph (see Figure 5.2). Each such replacement multiplies the number of proper 5-colourings by a factor of 480, as may be (somewhat tediously) checked.

The resulting graph after all these replacements have been carried out is 3-regular. $\quad\square$

5.10.2 Approximate Sampling

Consider the following *heat-bath* Markov chain on graph colourings: for a Markov chain in state X:

1. Pick a vertex v, uniformly at random.

2. We will re-colour v and all of its neighbours. Choose such a re-colouring uniformly at random from the set of all re-colourings that result in proper colourings of the graph. We will call this set of re-colourings $S_X(v)$.

A proof of rapid-mixing of this Markov chain may be obtained for 5-colouring graphs of maximal degree 3 and 7-colouring 4-regular graphs, using a non-trivial application of path coupling.

We will couple all pairs of states that have unit Hamming distance. The coupling procedure may be stated very simply, for X and Y two colourings that differ at only one vertex:

1. Pick a vertex v, uniformly at random.

2. We will re-colour v and all of its neighbours in both X and Y. Choose the joint distribution on $S_X(v)$ and $S_Y(v)$ such that the expected Hamming distance between the two re-coloured graphs is minimized.

The proof that this expected Hamming distance is always less than one is highly non-trivial, and involves the consideration of a very large number of cases. For each case it is necessary to show that in this case the expected Hamming distance is sufficiently less than one: this is equivalent to solving a large transportation problem for each case. The details of this proof may be found in Bubley, Dyer, and Greenhill [24]. The result, fortunately, may be simply stated: the mixing rate of this Markov chain is $O(n \log n \varepsilon^{-1})$ and the relaxation time is $O(n)$.

Chapter 6

Directions for Future Work

In this chapter we suggest a number of ideas and techniques that deserve further exploration and research, but are unfortunately outside the scope of this book.

6.1 Breaking Thresholds

In many of the problems that we have seen, we have encountered situations where we have an *fpras*, provided some parameter is greater than a threshold value, e.g. in graph k-colouring, we need in general to have $k \geq 2\Delta$, or in counting independent sets of size s, we need $s < n/2(\Delta + 1) + 1$. Such a result clearly begs for an explanation of such a threshold: are the problems on the other side of the threshold simply hard, maybe NP-hard? Or is the threshold chimeric, and merely an artefact of the analysis?

We will concentrate on the graph colouring problem that we examined in Sections 4.1 and 5.8. We saw a partial break of the threshold in a special class of graphs in Section 5.10, but here we examine the potential for a more generic break of the threshold.

The existence of a chromatic polynomial of a graph is well-known: its exact calculation is #P-hard, since knowledge of the chromatic polynomial gives the exact number of k-colourings for all k. Indeed discovery of much of the detail of the information about the chromatic polynomial remains #P-hard [7], however we know that we can gather certain information about it in randomized polynomial time, since we have an *fpras* for $k \geq 2\Delta$.

Specifically, suppose the chromatic polynomial of a graph on n vertices is $a_0 + a_1 x + a_2 x^2 + \cdots + a_n x^n$. (In fact we know some information about these coefficients *a priori* — see for example Stanley [116] for details.) Then an approximate solution to the number of colourings for any value of x may be considered simply as a pair of linear constraints on the variables $a_0, a_1, \ldots, a_n$. We may obtain such linear constraints (that are correct with sufficiently high probability) from our existing *fpras*. Then, answering the query "What bounds does this information give on the number of $2\Delta - 2$ colourings?"(say), is then simply a matter of solving two linear programs.

Of course this technique is unlikely to give us an *fpras* directly, but recall that we only need an approximation that is correct to within some polynomial factor in order to bootstrap this weak approximation to an *fpras*, using the techniques of Sinclair and Jerrum [112], and Jerrum, Valiant, and Vazirani [71] (see Section 3.3).

6.2 Beyond Self-Reducibility

Jerrum, Valiant, and Vazirani [71] and Sinclair and Jerrum [112] established an intimate connexion between approximate counting and approximate sampling for self-reducible structures. In practice, however, the power of these results is rarely, if ever used in applying the Markov chain Monte Carlo method.

Indeed there is good evidence to suppose that the right way to go forward is to have a weaker notion than self-reducibility to reduce approximate counting to approximate sampling. For instance, it is not known that volume approximation (or more generally log-concave integration) is self-reducible, yet this does not stop us from applying the Markov chain Monte Carlo method.

Establishing the right (weaker) condition under which approximate counting reduces to approximate sampling is an obvious area for future research[1]. Below we (informally) present a possible condition:

- There must be a partial order (*simplicity*) on the structures that we wish to be able to approximately count, such that:

  1. For every structure for which there is no simpler structure we can approximately count the size of the structure directly.

  2. For every structure for which there is a simpler structure we must be able (with high probability in randomized polynomial time) to find such a structure. Furthermore the ratio of sizes of these two structures must be bounded in expectation by a polynomial, and there must be a randomized polynomial-time recognizable embedding from one to the other.

  3. Thus given any starting structure we must be able to find a chain of successively simpler structures until our final structure, which we can approximately count directly. We must therefore insist that we can find (in polynomial time) such a chain with a polynomially bounded length.

It should be clear that if we can approximately sample from these structures then we can approximately count as well, i.e the existence of a fully-polynomial almost uniform sampler for such structures would imply the existence of an *fpras*.

---

[1]Dyer and Greenhill [47] have recently and independently arrived at a similar conclusion, and proposed a slightly different direction for future work.

6.3 Mixed Methods for Approximate Counting

We looked briefly at the Karp–Luby technique in Section 3.2.2, and saw that at its heart it required one to be able to sample and count (exactly) from a simpler set. Similarly, as we have seen throughout most of the rest of this book, the Markov chain Monte Carlo method also frequently requires one to sample from or count simpler sets. In both of these methods, it is sufficient to count and sample approximately. It would be interesting, therefore, to establish if there were any problems to which use of a mixed technique was advantageous, with say one method being used to bootstrap the other.

6.4 Faster Reductions from Approximate Counting to Approximate Sampling

In constructing a Markov chain Monte Carlo method, we have to choose a sequence of simpler structures, in order to calculate a set of ratios. Ultimately, we need to simplify the structure to a point where we can directly calculate or estimate our desired value for some "simplest" structure (see Section 6.2).

Frequently we will be faced with a choice of more than one notion of a "simpler" structure, e.g. if we are counting orientations of a graph with some particular property, we could choose to delete edges of the underlying graph until there were none left, or to add edges until we have a complete graph, or indeed to delete vertices until there were none left.

How do we choose between them? We present some heuristic arguments below; formalizing these arguments would require additional research.

For the sake of a heuristic argument, suppose we have an *a priori* bound on the ratio between our starting structure and our simplest structure, R, and that we can opt to find k intermediate structures with *a priori* ratio bounds $r_1, r_2, \ldots r_k$, such that $R = r_1 r_2 \ldots r_k$. Assume that the mixing rates of all of the Markov chains with state spaces these successively simpler structures are $\tau(\varepsilon)$, with relaxation times τ_2.

If we use the Markov chain Monte Carlo Lemma, then such an *fpras* takes time $O(\tau \sum_i (1 + r_i) k \varepsilon^2 \log k \varepsilon^{-1})$. In particular, if all the r_i are equal to $\sqrt[k]{R}$, this simplifies to $O(\tau(1 + \sqrt[k]{R}) k^2 \varepsilon^2 \log k \varepsilon^{-1})$. If we can choose k freely to minimize this expression, we discover we need $k = O(\log R)$, which would result in a time bound for the *fpras* of

$$O\left(\tau (\log R)^2 \varepsilon^2 \log(\varepsilon^{-1} \log R) \right).$$

Alternatively, using the method of Aldous (see Section 3.4), this may be achieved in time $O(k\tau + \sum_i (1 + r_i) k \varepsilon^{-2} \log(k) \tau_2)$. In particular, if all the r_i are equal to $\sqrt[k]{R}$, this simplifies to $O(k\tau) + O((1 + \sqrt[k]{R}) k^2 \log(k) \varepsilon^{-2} \tau_2)$. In a typical application, this will be dominated by the second term. If we can choose k freely to minimize this term, we discover we once again need $k = O(\log R)$, which would result in a time bound of

$$O\left(\tau \log R + \tau_2 (\log R)^2 \varepsilon^2 \log(\log R) \right).$$

These heuristic calculations suggest two things. Firstly, we should seek to ensure that our ratios r_i are bounded by a constant, rather than an arbitrary polynomial. Secondly, given the choice between a variety of "simplest" structures, we should opt for the one that would minimize $|\log R|$, where R is the *a priori* ratio bound.

6.5 Anti-ferromagnetic Models

We saw in Section 5.9 how the path coupling method may be applied to give a rapid-mixing result for the anti-ferromagnetic Potts model. Because of the nature of anti-ferromagnetic models, similar results should be obtainable through path coupling for other anti-ferromagnetic models. The following models are discussed in Simon [109]:

1. The *spin S Ising model*. Here each vertex takes on a "colour" in $\{0, 1, \ldots, S\}$. The Hamiltonian is given as the sum over all edges of the modulus of the difference of the values at its endpoints; i.e. for a colouring χ, and graph $G = (V, E)$, the Hamiltonian $\mathcal{H} = \sum_{(x,y) \in E} |\chi(x) - \chi(y)|$.

2. The *planar rotor* ($n = 2$), the *classical Heisenberg model* ($n = 3$) or, in general, the *n-vector model*. Here, each vertex takes on a "colour" with value a vector on the surface of the unit sphere in n dimensions. For a colouring χ, and graph $G = (V, E)$, the Hamiltonian is given by a sum of scalar products: $\mathcal{H} = \sum_{(x,y) \in E} \chi(x) \cdot \chi(y)$.

The second of these might prove to be a particularly interesting model to analyze, as it has a continuous state space. An appropriate random walk to consider is probably a Metropolis random walk, in which a new vertex and colouring vector are chosen uniformly at random, and then accepted or rejected based upon the Metropolis criterion. This could then be coupled in three stages: coupled so that the vertex being updated is the same, so that the vector chosen is the same, and so that the acceptance probabilities are coupled, so as to maximize the chance of the coupling copies accepting or rejecting simultaneously.

6.6 Log-Concave Sampling via Path Coupling

If we consider first the ease with which many proofs of rapid-mixing may be carried out with path coupling, and then consider the complexity of the analysis of a random walk for sampling from a log-concave distribution (see Section 4.3 and the references therein), it seems natural to attempt to marry the two.

Consider the following idea for analyzing a random walk similar to the "ball walk" used in Section 4.3 (we borrow much notation from this section too). Suppose X and Y are the states of two copies of the ball walk. We might couple X and Y in the following fashion: we consider a straight-line path from X to Y, where each element is $\alpha\sigma = \alpha r / \sqrt{n}$ from the next (there will be an end effect, since the path in general will not have length divisible by $\alpha\sigma$ but we will ignore this for the moment).

First some notation. Define

$$\phi(x) = \frac{1}{\sqrt{2\pi}}e^{-x^2/2}, \qquad \Phi(x) = \int_{-\infty}^{x} \phi(x')\mathrm{d}x', \qquad \Psi(x) = \int_{-\infty}^{x} x'\phi(x')\mathrm{d}x'.$$

Without loss of generality then, let us concentrate on points C and D at distance $\alpha\sigma$ from each other. Pick a point C' with the Gaussian distribution about C. We will pick D' in the same fashion as we did in Section 4.3.3.3 ("when the points are close together"). Then $Pr(C' = D') = 2\Phi(-\alpha/2)$.

If F is the log-concave distribution, we will further assume (without loss of generality) that $F(C) \le F(D)$. It may be more natural to use acceptance probabilities that are $M(a,b) = \min\{1, F(b)/F(a)\}$ — the "true" Metropolis condition.

Let $A_C = M(C, C')$, and $A_D = M(D, D')$. We couple the acceptances probabilities in the following way: choose w uniformly from $[0, 1]$. If $A_C \le w$, accept in C, and if $A_D \le w$ accept in D. We will write $\hat{C}$ and $\hat{D}$ for the points that C and D become after a step of the coupling.

To calculate the expected distance after one step, we first condition on whether $C' = D'$. Thus

$$\mathbf{E}\left(\mathrm{d}_{\mathrm{TV}}(\hat{C} - \hat{D},)\right) = 2\Phi\left(-\frac{\alpha}{2}\right)\mathbf{E}\left(\mathrm{d}_{\mathrm{TV}}(\hat{C} - \hat{D},|)C' = D'\right)$$
$$+ \left(1 - 2\Phi\left(-\frac{\alpha}{2}\right)\right)\mathbf{E}\left(\mathrm{d}_{\mathrm{TV}}(\hat{C} - \hat{D},|)C' \ne D'\right).$$

We will need to deal with these terms separately. Now

$$\mathbf{E}\left(\mathrm{d}_{\mathrm{TV}}(\hat{C} - \hat{D},|)C' = D'\right) =$$
$$\int_{\mathbb{R}^n} [\mathcal{L}(C'|C' = D')](x)\left[\begin{array}{l}||x - D||(M(C,x) - M(D,x)) \\ +\alpha\sigma(1 - M(C,x))\end{array}\right]\mathrm{d}x$$

and similarly

$$\mathbf{E}\left(\mathrm{d}_{\mathrm{TV}}(\hat{C} - \hat{D},|)C' \ne D'\right)$$
$$= \int_{\mathbb{R}^n} [\mathcal{L}(C'|C' \ne D')](x)\left[\begin{array}{l}||x - x'||\min\{M(C,x), M(D,x')\} \\ ||x - D||\max\{0, M(C,x) - M(D,x')\} \\ ||x' - C||\max\{0, M(D,x') - M(C,x)\} \\ +\alpha\sigma(1 - \max\{M(C,x), M(D,x')\})\end{array}\right]\mathrm{d}x$$

where x' is the point x reflected in the hyperplane of points equidistant from C and D.

Unfortunately, the analysis currently seems rather tricky. The difficulty over the analysis in Section 4.3 seems to stem from the need not only to consider the probability of coupling, as we do in Section 4.3.3.3, but also the need to consider the expected distance apart that the points move in the event of *not* coupling.

These difficulties will no doubt be overcome in the future; and then α may be picked near-optimally to give the best mixing rate.

Recall that we showed in Section 4.3 that the ball walk has mixing time $O^*(n^9)$, which may be simulated in $O^*(n^{6.5})$ super-steps. We will end this section with a conjecture.

CONJECTURE 4. *The mixing time for the ball walk is actually $\Theta(n^3 \log n)$.*

Appendix A

An Application of Dobrushin's Uniqueness Criterion

In this appendix, we explore an application of Dobrushin's uniqueness theorem to the problem of hypergraph colourings and the extended anti-ferromagnetic Potts framework. The result we obtain is weaker than that we obtained from path coupling (see Sections 5.8 and 5.9), but stronger than that obtained from coupling directly (Section 4.1).

We will use the notation from Theorem 2. We will consider the same Markov chain, $\mathcal{M}(G, k)$, as we did in Section 4.1; in statistical physics terminology, this is said to have the Metropolis kernel.

Suppose we have two colourings, X and Y, that differ only at vertex i. Then unless i and j are adjacent or equal, $\nu_{X,j}$ must be equal to $\nu_{Y,j}$, in which case $\varrho_{ij} = 0$.

If i and j are adjacent, then $d_{\mathrm{TV}}(\nu_{X,j}, \nu_{Y,j}) \leq 1/k$, since there will be at most one colour that i could be re-coloured to in X but not in Y.

For the case $i = j$, observe that $\nu_{X,j} > 0$ if and only if $\nu_{Y,j} > 0$, and in fact the only difference in the distributions is due to the chance of a step being rejected. Since this has a probability of at most Δ_1/k, this is an upper bound on ϱ_{jj} for all $j \in V$.

Now $\alpha = \max_{j \in V}\left\{\sum_{i \in V} \varrho_{ij}\right\}$, and since there are at most Δ_3 vertices adjacent to any given vertex, $\alpha \leq \Delta_3/k + \Delta_1/k$. To ensure that $\alpha < 1$, it is sufficient to have $k > \Delta_1 + \Delta_3$, and should this be the case, Theorem 2 applies and the result is established.

COROLLARY 16. *A Markov chain $\mathcal{N}$, satisfying Theorem 2 with stationary distribution the Gibbs distribution for the extended Potts framework, and the Metropolis kernel, is rapidly mixing, provided that $k > (1 - e^{\beta c_1})\Delta_1 + (1 - e^{\beta c_2})\Delta_3$, where $\alpha_1 = \max_{j \in V}\{\delta_2(j)/\delta_1(j)\}$, and α_2 is the maximum co-degree of G, i.e. the size of the largest set of edges that share at least two vertices. It is worth noting that both α_1 and α_2 are no greater than Δ_2, and that, consequently, when G is a graph, it is sufficient to take $k \geq 2(1 - e^{\beta})\Delta$, a bound that is observed by Jerrum [67].*

Proof. We have $\varrho_{ij} = 0$ unless i and j are either adjacent or equal. Consider then, two colourings, X and Y, where $Y = X_{i \to c}$ for some $c \in C$. We will assume, without loss of generality, that $H(X) \geq H(Y)$, and that $X \neq Y$.

Consider first the case where i and j are adjacent. We wish to re-colour vertex j. Suppose the colour choice is different from the colour of i in both X and Y. Then the probability of a colour choice being accepted is identical for both X and Y. If we had instead chosen the colour that i holds in X, then the probability of acceptance would be $\min\{1, e^{-\beta(f+g)}\}$ for X, and $\min\{1, e^{-\beta(f-h)}\}$ for Y, where f is the difference in the Hamiltonians that is due to edges not including both i and j, g is the number of edges containing i that are critical for j in X, and h the number of edges containing i that are critical for j in Y. Observe that $g + h \leq \alpha_2$.

Note that if $f \leq -g$, then the difference in probability of a colour being accepted is nil, so we may assume in what follows that $f \geq 1 - g$. Thus the difference in probabilities is bounded by $\min\{1 - e^{-\beta(f+g)}, e^{-\beta(f-h)}(1 - e^{-\beta(g+h)})\}$. Suppose $f \leq h$: then $1 - e^{-\beta(f+g)} \leq 1 - e^{-\beta\alpha_2}$. Suppose instead that $f \geq h + 1$: then $e^{-\beta(f-h)}(1 - e^{-\beta(g+h)}) \leq 1 - e^{-\beta\alpha_2}$ too, whence we obtain that $\varrho_{ij} \leq (1 - e^{-\beta\alpha_2})/k$.

For the case of ϱ_{jj}, we will assume once more that $H(X) \geq H(Y)$, and that $X \neq Y$. Then ϱ_{jj} is bounded above by the probability of the colour choice made from colouring Y being rejected. Now there can be at most $\delta_1(j)$ colours that have a non-zero probability of rejection; enumerate these by $c_1, c_2, \ldots, c_r$. Then the probability of rejection is bounded above by $\sum_{l=1}^{r}(1 - e^{-\beta\sigma_l})$, where σ_l is the number of edges critical for j with colour c_l. Now $\sum_{l=1}^{r}\sigma_l \leq \delta_2(j)$, so $\varrho_{jj} \leq \sum_{l=1}^{r}(1 - e^{-\beta\sigma_l})/k \leq \delta_1(j)[1 - e^{-\beta\delta_2(j)/\delta_1(j)}]/k \leq \Delta_2[1 - e^{-\beta\alpha_1}]/k$.

Summing the ϱ_{ij} completes the proof.　　　　$\square$

Appendix B

A Hierarchy of #SAT Restrictions

B.1 Introduction

The problem of SAT is without a doubt the most famous problem in the field of computational complexity. Since Cook [30] first introduced it, it has been looked at extensively in a diverse range of areas and in a variety of guises. Since our primary concern in this book is with counting problems, we turn naturally to the problem of #SAT.

Since, in its full generality, #SAT is hard even to approximate [111], we look at a variety of natural ways of restricting the problem. The choice of which restrictions to look at is necessarily arbitrary, but we choose the restrictions below because we may then restate a variety of results in a convenient form.

The linchpin problem is of course #SAT itself, which, when given a Boolean formula in conjunctive normal form returns the number of satisfying assignments to the variables in the formula.

We restrict the classes of formulæ to be considered in three hierarchical ways.

1. We consider restricting the type of clause to be Horn (HORN), i.e. each clause contains at most one unnegated variable, or to be monotone (MON), i.e. every literal is unnegated.

   Remark. Note that by negating every variable in a formula, we parsimoniously transform MON problems to ANTI-MON problems in which every literal is negated. Thus ANTI-MON is a restriction of HORN which in turn is a restriction of SAT. For the sake of brevity, we shall henceforth refer to MON as if it were itself a restriction of HORN. ○

2. We consider restricting the number of literals in each clause to be at most k for some natural number k. We write this as #kSAT (or #kHORN, or #kMON, respectively). Note that #kSAT is used by some authors (e.g. Garey and Johnson [51]) to mean each clause contains *exactly* k literals. We use the relaxed definition of *at most* k literals to ensure that we have a hierarchy, where #kSAT is a restriction of #$(k + 1)$SAT.

3. We consider restricting the number of occurrences of each variable in the entire formula to be at most l for some natural number l. We write this as #$l\mu$-SAT. This notation is also used by Roth [105] and Vadhan [122]. This is sometimes referred to elsewhere in the literature as READ-l-SAT. In Section 4.2 and in [20], #2μ-SAT was referred to as TWICE-SAT.

As we observed in Section 4.2, there is a direct correspondence between #2μ-SAT and the problem of finding sink-free orientations in a graph, where each edge in the graph can be one of two types: ordinary or skew, and each type of edge has two orientations. Furthermore, #2μ-MON may be recast as #EDGE-COVERINGS, which counts all sets of edges in a graph which cover all vertices.

Then #2μ-kSAT (or #2μ-kMON, respectively) is the restriction of these graphical problems to the case where the degree of the graph is at most k.

As a dual interpretation, we may regard #2MON as the problem #INDEPENDENT-SETS, which counts all independent sets in a graph. The reduction here is to associate each variable in the Boolean formula with a vertex in the graph, and each clause with an edge (a clause containing literals a and b is equivalent to an edge between vertices a and b). Then every satisfying assignment of the Boolean formula corresponds to an independent set in the graph: the exact correspondence being between vertices in the independent set and variables taking value False. This reduction is due to Sinclair [111].

The restriction #$l\mu$-2MON may then be interpreted graphically as counting independent sets in a graph of degree at most l.

Of course there are also graphical interpretations when we move from #2MON and #2μ-SAT to kMON and #$l\mu$-SAT: we just move from graphs to hypergraphs, adjusting the interpretation as necessary. Thus #$l\mu$-kMON may be interpreted as counting the number of sink-free orientations of a hypergraph, where each edge contains at most l vertices, and each vertex is present in at most k edges. Each edge may then be oriented in one of two ways, such that either all vertices are pointed towards, or no vertices are. Equivalently, this is the hypergraph equivalent of #EDGE-COVER, where one counts the number of sets of edges that cover all vertices (with k and l having the same interpretation as for the hypergraph version of #SFO).

Alternatively, we may interpret #$l\mu$-kMON as counting the number of independent sets in a hypergraph where all edges contain at most k vertices, and each vertex is present in at most l edges. Here, an independent set is any set of vertices which is not a superset of any edge.

The interpretation as sink-free hypergraph orientations extends readily to the cases of #$l\mu$-kHORN and #$l\mu$-kSAT. Once again, we consider hypergraphs where each edge contains at most l vertices, and each vertex is present in at most k edges. Each edge may then be oriented in one of two ways. The two ways complement each other in the sense that vertices that are pointed to in one orientation are not pointed to in the other and vice versa. In the MON case, one of the ways is always such that no vertices were pointed to. In the HORN case, one of the ways is such that at most one vertex is pointed to, and in the SAT case there is no restriction at all.

B.2 A Summary of Known Results

There are essentially four classes of results that we will consider:

1. Results which prove that exact counting is possible in polynomial time.

2. Results which prove that the problem of counting exactly is #P-complete.

3. Results which prove that approximate counting is possible in randomized polynomial time (i.e. there is an *fpras*).

4. Results which prove that approximate counting is NP-hard (i.e. that no randomized polynomial-time algorithm exists for approximate counting unless $NP \subseteq BPP$).

B.2.1 Easy Exact Counting

Unsurprisingly, there are very few problems in our hierarchy for which exact counting is known to be possible in polynomial time.

There are two relevant results. The first is due to Vadhan [122], who shows that 2μ-2MON may be solved using a simple recurrence formula. The second is due to Roth [105], who generalizes this result to 2μ-2SAT. His algorithm may be simplified substantially by regarding the problem as a problem on sink-free graph orientations. We present this argument in the next section.

B.2.1.1 A simple polynomial-time algorithm for #2μ-2SAT

We interpret the problem of #2μ-2SAT as counting sink-free orientations of a graph of degree at most 2.

Firstly, split the graph up into connected components; we will multiply the number of sink-free orientations of each component together for the final answer.

Secondly, we allow a connected component to have one distinguished "edge". This edge can take each of the four orientations (two each associated with ordinary and skew edges) with some multiplicity.

We proceed by recursion. The input is a connected component with one edge replaced by this distinguished edge. The base cases of the recursion are those for which there is only one (distinguished) edge. These are simple to deal with: if the graph is a self loop, then return the sum of the multiplicities of the distinguished edges, save the (skew) orientation that points outwards; if the graph is a path of length one, then return the multiplicity of the (skew) orientation that points inwards.

If we have a case that is not a base case, then this too is simple to deal with: consider the distinguished edge, and an edge adjacent to it. We will replace these two edges by a single distinguished edge, and this new graph will be dealt with by recursion.

The multiplicities of this new distinguished edge are calculated by considering all possible pairings of orientations of the two edges that do not create a sink between them. So, for example, supposed initially the distinguished edge can point inwards with multiplicity a, outwards with multiplicity b, right with multiplicity c and left with multiplicity d, and the edge that we are considering that is adjacent to it is on its right and is an ordinary edge. Then the new distinguished edge will point inwards with multiplicity a, outwards with multiplicity $b + d$, right with multiplicity $a + c$ and left with multiplicity d.

B.2.2   Hard Exact Counting

There have been many #P-completeness results that may be fitted within our #SAT hierarchy, starting with one of Valiant's original #P-completeness results [125].

Roth [105] showed that by using a simple rewriting technique, one could effect a parsimonious reduction from #2MON to #3μ-2HORN, and thus this latter problem is #P-complete too.

Vadhan [122] showed that #4μ-2MON is #P-complete; he later strengthened this result in the setting of independent sets, to show that counting independent sets in planar bipartite graphs of maximal degree 4 was #P-complete too [123].

Dyer and Greenhill [44] recently strengthened Vadhan's result to show that #3μ-2MON is #P-complete.

Finally, Bubley and Dyer [20] showed that #2μ-SAT is #P-complete and mentioned that #2μ-MON was too. Both of these results are discussed in Section 4.2.3.3. Both of these results are established by a non-parsimonious reduction (polynomial interpolation), ultimately from perfect matchings.

B.2.3   Easy Approximate Counting

There are few results in this area. All of the results mentioned here rely on the Markov chain Monte Carlo method, and in particular rely on coupling techniques for their proof of rapid mixing.

Dyer and the author [20] showed that #2μ-SAT has an *fpras*. This analysis is recapitulated in Section 4.2. The complexity of the *fpras* was subsequently improved by a tighter analysis of the Markov chain using path coupling [21] (see also Section 5.2).

Luby and Vigoda [91] showed that #4μ-2MON has an *fpras*, again using a coupling argument. Dyer and Greenhill [43] subsequently improved the complexity of the *fpras* by using a different Markov chain for this problem, and analyzing this new Markov chain by path coupling.

Recently, Dyer and Greenhill [43] have given a rapidly mixing Markov chain for #3μ-3MON, once again utilizing path coupling.

B.2.4   Hard Approximate Counting

Again, very few results are known about the hardness of approximately counting problems in our chosen framework.

Sinclair [111] shows that the existence of an *fpras* for #2MON would imply that $RP = NP$, by showing that such an fpras could be used to construct a randomized polynomial-time algorithm that could answer the NP-complete query "Does the input graph contain an independent set of size k?" [51].

Roth's parsimonious reduction from 2MON to #3μ-2HORN, which we mentioned in Section B.2.2, immediately powers Sinclair's result on the inapproximability of #2MON to the inapproximability of #3μ-2HORN .

Only one other hardness of approximation result is known for our hierarchy; we go into details in the next section.

B.2.4.1 An NP-hardness proof for approximating #$k\mu$-2MON

In this section, we show the existence of *some* natural number k, for which #$k\mu$-2MON is inapproximable, unless NP $\subseteq$ BPP. The technique used here is based on an idea of Dyer. Dyer, Frieze, and Jerrum [41] have recently shown that approximating #$K\mu$-2MON is hard for $k \geq 25$.

Recent results on the hardness of approximation of some optimization problems have ramifications for approximate counting. Arora *et al.* [10] showed that for every MAX SNP-hard problem, there is some $\varepsilon > 0$ for which finding an approximation within ratio $1 + \varepsilon$ is NP-hard. Berman and Fujito [16] showed that the problem of finding the maximum independent set in graphs of degree 3 is MAX SNP-complete. Alon *et al.* [5] show that there is a constant, $\varepsilon > 0$, such that approximating the size of the maximum independent set in graphs of degree Δ is NP-hard within ratio Δ^ε, for every $\Delta \geq 3$, and thus the problem of finding the maximum independent set in graphs of degree Δ is MAX SNP-complete for every $\Delta \geq 3$.

Given a graph $G = (V, E)$ on n vertices of maximum degree Δ, construct another graph G' in the following manner. Let k be a natural number (to be determined later). Replace each vertex v in G with a set of k vertices, S_v. If (a, b) is an edge of G, then form edges in G' between all vertices in S_a and S_b. Note that since the maximum degree of G is at most Δ, the maximum degree of G' is at most $k\Delta$.

We shall proceed by contradiction. Let m be the size of the largest independent set in G, a graph of degree Δ. We shall use an RP algorithm to try all values successively from (say) n downwards until we find an approximant for m, within factor $1/\alpha$. Note that trivially, $n/(\Delta+1) \leq m \leq n\Delta/(\Delta+1)$, and so we will only seek an approximant within those bounds.

Suppose we are considering $\hat{m}$, having established (with high probability) that there is no independent set of size $\hat{m} + 1$. If there is an independent set of size $\hat{m}$ in G, then there must be an independent set of size $k\hat{m}$ in G' (i.e. all the vertices in all the sets S_v that are formed from the independent set in G). Since a subset of an independent set is itself an independent set there would thus be at least $2^{k\hat{m}}$ independent sets in G'. If, on the other hand, there is no independent set of size $\lceil \alpha\hat{m} \rceil$, then the number of independent sets of G' is at most

$$\sum_{i=0}^{\lceil \alpha\hat{m} \rceil - 1} \binom{n}{i} 2^{(k-1)i} \leq 2^{n+(k-1)\alpha\hat{m}}.$$

Thus, if

$$(1 + \varepsilon)2^{n+(k-1)\alpha\hat{m}} < 2^{k\hat{m}}$$

then we can, with high probability tell if $\hat{m}$ will suffice as an approximant to m.

It is sufficient, therefore, to take

$$k > \frac{\log_2(1+\varepsilon) + n - \alpha\hat{m}}{\hat{m}(1-\alpha)}.$$

Recall that $\hat{m} > n/(\Delta+1)$. Thus, we need only take

$$k > 1 + \frac{\Delta}{1-\alpha} + \frac{(\Delta+1)\log_2(1+\varepsilon)}{n(1-\alpha)}.$$

We may clearly neglect the last term, since we may insist n is sufficiently large that the term is less than $k - 1 - \Delta/(1-\alpha)$.

To summarize, then: if it is hard to approximate the maximum size of an independent set in graphs of degree Δ within factor $1/\alpha$, then there is no *fpras* for counting all independent sets in graphs of degree $\Delta\lceil 1 + \Delta/(1-\alpha)\rceil$.

B.3   Summary and Conclusions

We have listed all of the known results on easy and hard counting, both exact and approximate, for all problems within our hierarchy. Our knowledge is by no means complete however, and as we see in Figures B.1, B.2, and B.3, there remain many open problems.

The weakest of the remaining open problems are:

1. Is #2μ-3MON polynomial-time computable?

2. Is #2μ-kSAT #P-complete for some constant k?

3. Is there an *fpras* for #5μ-2MON, #4μ-3MON, or #3μ-4MON?

4. Is #24μMON hard to approximate?

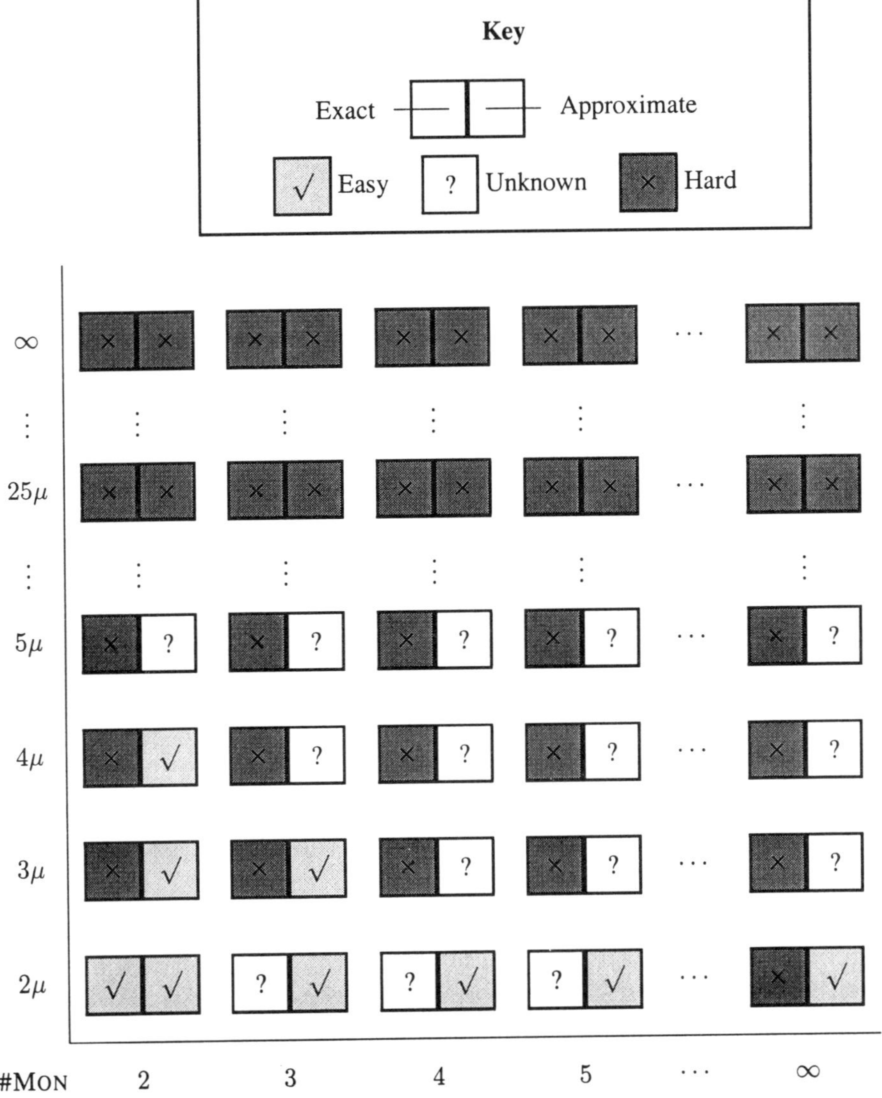

Figure B.1: A hierarchy of #MON restrictions.

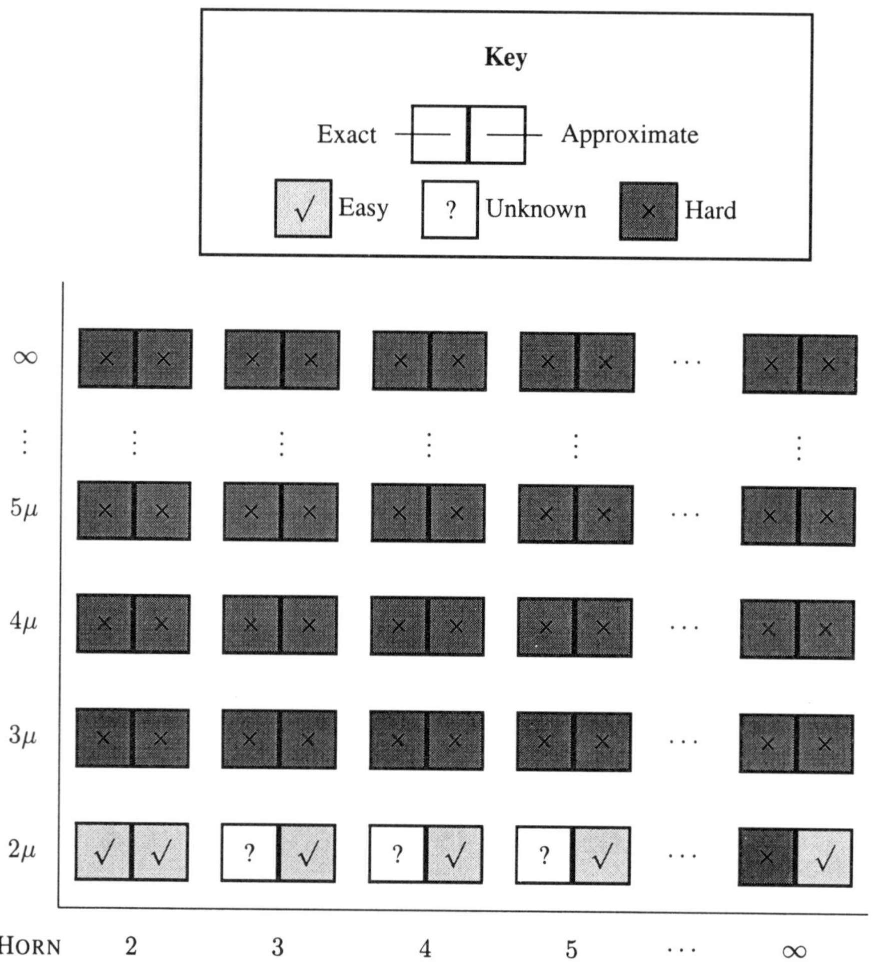

Figure B.2: A hierarchy of #HORN restrictions.

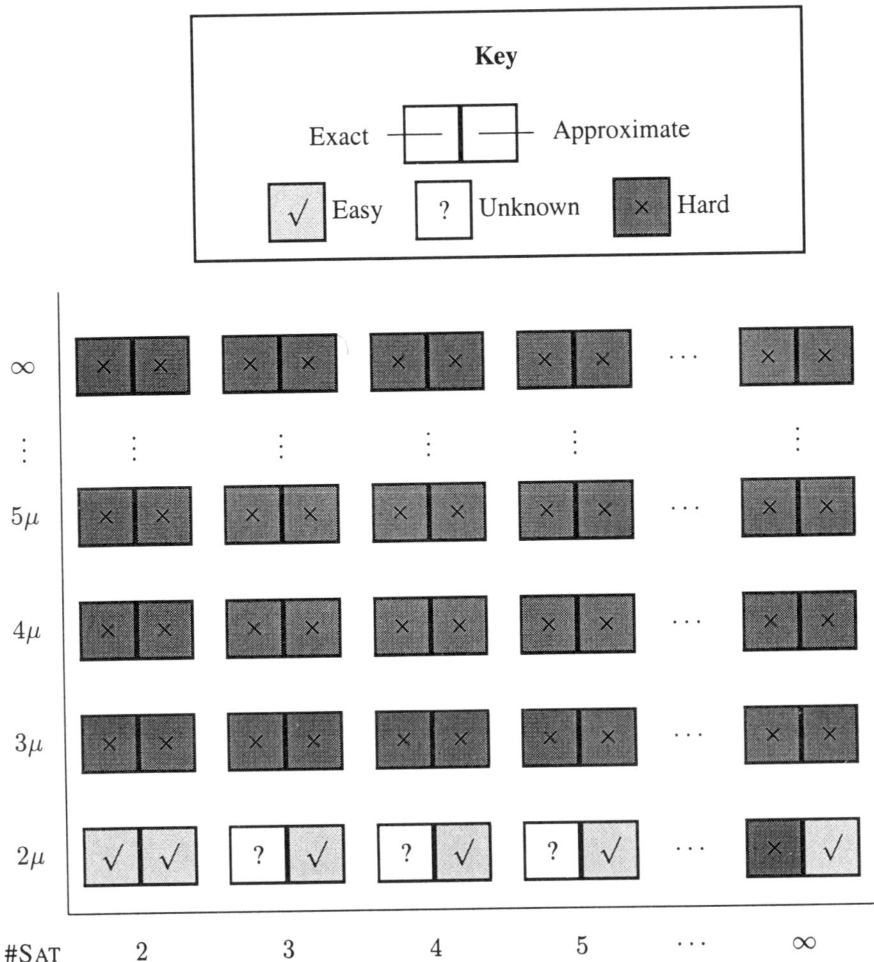

Figure B.3: A hierarchy of #SAT restrictions.

Appendix C

Equivalence of Transposition Distance to Spearman's Footrule

In this appendix we consider two metrics on the set of linear extensions of a partial order, and show that these metrics are equivalent. We made use of this fact in Section 5.7.

When we refer to a total order (or permutation) we take this to be implicitly on the set $\{1, 2, \ldots, n\}$. Furthermore, if X is a permutation, then $X(i)$ denotes the position that i is moved to under the permutation. Thus the identity permutation, I, could be written as $I(i) = i$. When we compose permutations, we mean $XY(i) = X(Y(i))$.

We defined *Spearman's footrule* as a metric on total orders:

$$\delta_S(X, Y) = \tfrac{1}{2} \sum_{i=1}^{n} |X(i) - Y(i)|.$$

It should be noted that this definition of Spearman's footrule is in accord with Spearman's usage [114] and that recommended by [39]; other authors (e.g. [34]) drop the half from the definition.

THEOREM 7. *Suppose X and Y are distinct total orders. Suppose further that they differ by more than a single transposition (i.e. there is no transposition T, such that $X = TY$), and that both X and Y are linear extensions of a partial order, P. Then there exists a transposition T' such that $X' = T'X$ is a linear extension of P distinct from X and Y, and $\delta_S(Y, X') + \delta_S(X', X) = \delta_S(Y, X)$.*

Proof. We may assume, without loss of generality (by relabelling, say), that $Y = I$. Suppose we have i and j, such that $i \leq X(j) < X(i) \leq j$. Recall that one can write a permutation as the product of disjoint cycles. In any cycle we may find such an i and j. If we have the further condition that for all k such that $X(j) < k < X(i)$, $X(k) = k$, then we are done, for we may take $X' = \sigma(X(i), X(j))X$.

Suppose that we do not have this condition. Then either there is an entire cycle with all its elements between $X(j)$ and $X(i)$, in which case we may start again with this cycle, or there is either an i' such that $i' < X(j)$ or $i' > X(i)$, and $X(j) < X(i') < X(i)$, in which case we may start again with one of i and i' or j and i'.

This completes the proof by the classical method of infinite descent. $\qquad\square$

COROLLARY 17. *Spearman's footrule is identical to transposition distance.*

Proof. Let X and Y be arbitrary linear extensions of a partial order. If $X = Y$, or X and Y differ by exactly one transposition, then the equivalence of the two metrics follows easily. If, instead, X and Y differ by more than a single transposition, then we may proceed by repeated application of the above Theorem. We may use the Theorem to construct a transposition sequence. The minimality of δ then guarantees that $\delta \leq \delta_S$. Conversely, we may argue by induction on the length of the minimum transposition sequence. If $Z \neq X, Y$ is a member of this sequence, then

$$\delta(Y, X) = \delta(Y, Z) + \delta(Z, X) = \delta_S(Y, Z) + \delta_S(Z, X) \geq \delta_S(Y, X)$$

respectively by induction and the triangle inequality. Thus $\delta_S \leq \delta$. $\qquad\square$

Bibliography

[1] Aldous D. Random walks on finite groups and rapidly mixing Markov chains. In: Dold A, Eckmann B, editors, Séminaire de Probabilités XVII 1981/1982, New York: Springer-Verlag, volume 986 of Springer-Verlag Lecture Notes in Mathematics. 1983; 243–297. 8, 15, 16

[2] Aldous D. On the Markov chain simulation method for uniform combinatorial distributions and simulated annealing. Probab Engrg Inform Sci 1987;1:33–46. 16, 36

[3] Aldous D, Fill JA. Reversible Markov chains and random walks on graphs, October 1994. Chapter 4. Monograph in preparation. 8, 16, 36, 111

[4] Alon N. Eigenvalues and expanders. Combinatorica 1986;6(2):83–96. 23

[5] Alon N, Feige U, Wigderson A, Zuckerman D. Derandomized graph products. Computat Complexity 1995;5(1):60–75. 131

[6] Alon N, Frieze A, Welsh D. Polynomial time randomized approximation schemes for Tutte–Gröthendieck invariants: The dense case. Random Structures Algorithms 1995;6(4):459–478. 10, 14, 32, 49, 100

[7] Annan JD. The complexities of the coefficients of the Tutte polynomial. Discrete Appl Math 1995;57(2–3):99–103. 119

[8] Applegate D, Kannan R. Sampling and integration of near log-concave functions. In: Proceedings of the Twenty Third Annual ACM Symposium on Theory of Computing. New Orleans, Louisiana, 6–8 May 1991; 156–163. 24, 27, 62, 63

[9] Arora S. The approximability of NP-hard problems. In: Proceedings of the Thirtieth Annual ACM Symposium on Theory of Computing. Dallas, Texas, 24–26 May 1998; 337–348. viii

[10] Arora S, Lund C, Motwani R, Sudan M, Szegedy M. Proof verification and hardness of approximation problems. In: Proceedings of the 33rd Annual Symposium on Foundations of Computer Science. Pittsburgh, Pennsylvania: IEEE, 24–27 October 1992; 14–23. 131

[11] Ausiello G, Crescenzi P, Gambosi G, Kann V, Marchetti Spaccamela A, Protasi M. Approximate Solutions of NP-Hard Optimization Problems. Springer-Verlag, August 1999. ISBN 3-540-65431-3. viii

[12] Barbour AD, Holst L, Janson S. Poisson Approximation. Number 2 in Oxford Studies in Probability. Oxford: Clarendon Press, 1992. ISBN 0-19-852235-5. 18, 66

[13] Ben-Dor A, Halevi S. 0–1 permanent is #P-complete, a simpler proof. In: Proceedings of the 2nd Israeli Symposium on Theoretical Computer Science. Netanya, Israel, 1993; 108–117. 5

[14] Berge C. Nombres de coloration de l'hypergraphe h-parti complet. In: Berge C, Ray-Chaudhuri DK, editors, Hypergraph Seminar. Springer-Verlag, Number 411 in Lecture Notes in Mathematics, 1972; 13–20. 38

[15] Berge C. Graphs and Hypergraphs. Amsterdam, London: North-Holland Publishing Company, volume 6 of North-Holland Mathematical Library. 1973; chapter 19, 428–447. 11, 39

[16] Berman P, Fujito T. On approximation properties of the independent set problem in degree 3 graphs. In: Akl SG, Dehne F, Sack JR, Santoro N, editors, Proceedings of the 4th International Workshop on Algorithms and Data Structures (WADS'95), Berlin: Springer-Verlag, volume 955 of Lecture Notes in Computer Science. ISBN 3-540-60220-8, 1995; 449–460. 131

[17] Brightwell G, Winkler P. Counting linear extensions. Order 1991;8(3):225–242. 105, 111

[18] Broder AZ. How hard is it to marry at random? (On the approximation of the permanent). In: 18th Annual ACM Symposium on Theory of Computing. Berkeley, California, 28–30 May 1986; 50–58. Erratum in *Proceedings of 20th Annual ACM Symposium on Theory of Computing*, p.551, 1988, Chicago, Illinois, 2–4 May 1988. 17

[19] Bubley R, Dyer M. Faster random generation of linear extensions. Technical Report 97.41, School of Computer Studies, University of Leeds, August 1997. 105

[20] Bubley R, Dyer M. Graph orientations with no sink and an approximation for a hard case of #SAT. In: Proceedings of the Eighth Annual ACM–SIAM Symposium on Discrete Algorithms. New Orleans, Louisiana, 5–7 January 1997; 248–257. 17, 48, 91, 128, 130

[21] Bubley R, Dyer M. Path coupling: A technique for proving rapid mixing in Markov chains. In: 38th Annual Symposium on Foundations of Computer Science. Miami Beach, Florida: IEEE. ISBN 0-8186-8197-7, 20–22 October 1997; 223–231. 83, 93, 96, 102, 115, 130

[22] Bubley R, Dyer M. Faster random generation of linear extensions. In: Proceedings of the Ninth Annual ACM–SIAM Symposium on Discrete Algorithms. San Francisco, California, 25–27 January 1998; 350–354. 105, 109, 110

[23] Bubley R, Dyer M. Faster random generation of linear extensions. Discrete Math 1999;201:81–88. 83, 105

[24] Bubley R, Dyer M, Greenhill C. Beating the 2Δ bound for approximately counting colourings: A computer-assisted proof of rapid mixing. In: Proceedings of the Ninth Annual ACM–SIAM Symposium on Discrete Algorithms. San Francisco, California, 25–27 January 1998; 355–363. 118

[25] Bubley R, Dyer M, Greenhill C, Jerrum M. On approximately counting colourings of small degree graphs. SIAM J Comput 1999;29:387–400. 83, 99, 115

[26] Bubley R, Dyer M, Jerrum M. An elementary analysis of a procedure for sampling points in a convex body. Random Structures Algorithms 1998; 12:213–235. 17, 62

[27] Bubley R, Dyer ME. Path coupling, Dobrushin uniqueness, and approximate counting. Technical Report 97.04, School of Computer Studies, University of Leeds, January 1997. 18, 83, 114

[28] Cai Jy, Hemachandra LA. Exact counting is as easy as approximate counting. Technical Report TR 86-791, Department of Computer Science, Cornell University, June 1986. 30

[29] Chung FRK. Logarithmic Sobolev techniques for random walks on graphs. http://math.ucsd.edu/~fan/. 26

[30] Cook SA. The complexity of theorem-proving procedures. In: Conference Record of Third Annual ACM Symposium on Theory of Computing. Shaker Heights, Ohio, 3–5 May 1971; 151–158. 4, 5, 127

[31] Cooper C, Frieze AM. Mixing properties of the Swendsen–Wang process on classes of graphs, January 1998. (Manuscript). 83

[32] Czumaj A, Kanarek P, Kutylowski M, Lorys K. Delayed path coupling and generating random permutations via distributed stochastic processes. In: Proceedings of the Tenth Annual ACM–SIAM Symposium on Discrete Algorithms. Baltimore, Maryland, 17–19 January 1999; 271–280. 83

[33] Dagum P, Luby M. Approximating the permanent of graphs with large factors. Theoret Comput Sci 1992;102:283–305. 55

[34] Diaconis P, Graham RL. Spearman's footrule as a measure of disarray. J Roy Statist Soc Ser B 1977;39(2):262–268. 107, 137

[35] Diaconis P, Saloff-Coste L. Comparison theorems for reversible Markov chains. Ann Appl Probab 1993;3(3):696–730. 24, 109

[36] Diaconis P, Saloff-Coste L. Logarithmic Sobolev inequalities and finite Markov chains. Ann Appl Probab 1996;6(3):695–750. 24, 26, 27

[37] Diaconis P, Shahshahani M. Generating a random permutation with random transpositions. Z Wahrsch Verw Gebiete 1981;57(2):159–179. 110

[38] Diaconis P, Stroock D. Geometric bounds for eigenvalues of Markov chains. Ann Appl Probab 1991;1(1):36–61. 22, 23, 25

[39] Dinneen LC, Blakesley BC. Definition of Spearman's footrule. J Roy Statist Soc Ser C 1982;31(1):66. 137

[40] Dyer M, Frieze A. Computing the volume of convex bodies: A case where randomness provably helps. In: Proceedings of the 44th Symposium in Applied Mathematics, American Mathematical Society, volume 44 of Symposia in Applied Mathematics. 1991; 123–169. 24, 29, 30, 62, 105, 106, 110, 111

[41] Dyer M, Frieze A, Jerrum M. On counting independent sets in sparse graphs. Technical report, Department of Computer Science, University of Edinburgh, July 1998. Report ECS-LFCS-98-391. 131

[42] Dyer M, Frieze A, Kannan R. A random polynomial-time algorithm for approximating the volume of convex bodies. J Assoc Comput Mach January 1991;38(1):1–17. 24, 62, 63, 105

[43] Dyer M, Greenhill C. On Markov chains for independent sets, December 1997. (Preprint), http://www.scs.leeds.ac.uk/rand/psfiles/ ind-sets.ps. 8, 25, 83, 103, 104, 105, 109, 110, 115, 130

[44] Dyer M, Greenhill C. Some #P-completeness proofs for colourings and independent sets. Technical Report 97.47, School of Computer Studies, University of Leeds, December 1997. 38, 130

[45] Dyer M, Greenhill C. A genuinely polynomial-time algorithm for sampling two-rowed contingency tables. In: Proceedings of the 25th International Colloquium on Automata, Languages and Programming. Aalborg, Denmark: EATCS, March 1998; 339–350. 83

[46] Dyer M, Greenhill C. A more rapidly mixing Markov chain for graph colourings. Random Structures Algorithms 1998;13:285–317. 83, 115

[47] Dyer M, Greenhill C. Random walks on combinatorial objects. In: Lamb JD, Preece DA, editors, Surveys in Combinatorics 1999, Cambridge: Cambridge University Press, volume 267 of London Mathematical Society Lecture Note Series. 1999; 101–136. 83, 120

[48] Felsner S, Wernisch L. Markov chains for linear extensions, the two-dimensional case. In: Proceedings of the Eighth Annual ACM–SIAM Symposium on Discrete Algorithms. New Orleans, Louisiana, 5–7 January 1997; 239–247. 105

[49] Frieze A, Kannan R. Log-Sobolev inequalities for sampling from log-concave distributions, July 1997. http://www.math.cmu.edu/~aflp/logsob.ps.gz. 27, 62

[50] Frieze A, Kannan R, Polson N. Sampling from log-concave distributions. Ann Appl Probab 1994;4(3):812–837. 6, 24, 25, 27, 62

[51] Garey MR, Johnson DS. Computers and Intractability: a Guide to the Theory of NP-Completeness. San Francisco, California: W. H. Freeman and Company, 1979. ISBN 0-7167-1045-5. 1, 2, 55, 127, 130

[52] Garey MR, Johnson DS, Stockmeyer L. Some simplified NP-complete graph problems. Theoret Comput Sci 1976;1:237–267. 5, 48, 55

[53] Georgii HO. Gibbs Measures and Phase Transitions. volume 9 of de Gruyter Studies in Mathematics. Berlin, New York: de Gruyter, 1988. ISBN 0-89925-462-4 or 3-11-010455-5. 18

[54] Gilbarg D, Trudinger NS. Elliptic Partial Differential Equations of Second Order. Springer-Verlag. Number 224 in Grundlehren der mathematischen Wissenschaften. ISBN 3-540-08007-4 or 0-387-08007-4, 1977; section 7.2. 76

[55] Goldberg LA. Efficient Algorithms for Listing Combinatorial Structures. Distinguished Dissertations in Computer Science. Cambridge University Press, 1993. ISBN 0-521-45021-7. 53, 94

[56] Gore V, Jerrum M, Kannan S, Sweedyk Z, Mahaney S. A quasi-polynomial-time algorithm for sampling words from a context-free language. Inform and Comp 10 April 1997;134(1):59–74. 32, 33

[57] Griffeath D. Coupling methods for Markov processes. In: Rota GC, editor, Studies in Probability and Ergodic Theory, Academic Press, volume 2 of Advances in Mathematics Supplementary Studies. 1978; 1–43. 70

[58] Grimmett GR, Stirzaker DR. Probability and Random Processes. Oxford University Press, 2nd edition, 1992. 6, 7

[59] Gross L. Decay of correlations in classical lattice models at high temperature. Comm Math Phys 1979;68:9–27. 18

[60] Grötschel M, Lovász L, Schrijver A. Geometric Algorithms and Combinatorial Optimization. New York: Springer-Verlag, 1988. 64

[61] Häggström O, Nelander K. Exact sampling from anti-monotone systems. Statistica Neerlandica 1998;52:360–380. 28

[62] Häggström O, Nelander K. On exact simulation of Markov random fields using coupling from the past. Scand J Statist 1999;26(3):395–411. 115

[63] Hoeffding W. Probability inequalities for sums of bounded random variables. J Amer Statist Assoc 1963;58(301):13–30. 73

[64] Huber M. Exact sampling and approximate counting techniques. In: Proceedings of the Thirtieth Annual ACM Symposium on Theory of Computing. Dallas, Texas, 24–26 May 1998; 31–40. 28, 48

[65] Jaeger F, Vertigan DL, Welsh DJA. On the computational complexity of the Jones and Tutte polynomials. Math Proc Cambridge Philos Soc 1990;108:35–53. 5, 100, 116

[66] Jerrum M. The "Markov chain Monte Carlo" method: Analytical techniques and applications. In: Proceedings of the IMA Conference on Complex Stochastic Systems and Engineering. Oxford University Press, 1995; 191–207. viii, 23, 24, 26

[67] Jerrum M. A very simple algorithm for estimating the number of k-colorings of a low-degree graph. Random Structures Algorithms 1995;7(2):157–165. 17, 18, 37, 38, 40, 111, 112, 113, 114, 125

[68] Jerrum M, Sinclair A. Approximating the permanent. SIAM J Comput December 1989;18(6):1149–1178. 17, 23

[69] Jerrum M, Sinclair A. Polynomial time approximation algorithms for the Ising model. SIAM J Comput 1993;22:1087–1116. 23

[70] Jerrum M, Sinclair A. The Markov chain Monte Carlo method: an approach to approximate counting and integration. In: Hochbaum D, editor, Approximation Algorithms for NP-Hard Problems, Boston: PWS Publishing. 1996; 482–520. 23, 24, 111

[71] Jerrum M, Valiant L, Vazirani V. Random generation of combinatorial structures from a uniform distribution. Theoret Comput Sci 1986;43:169–188. 1, 14, 30, 33, 56, 120

[72] Johnson DS, Yannakakis M, Papadimitriou CH. On generating all maximal independent sets. Inform Process Lett 1988;27:119–123. 52, 94

[73] Kannan R. Markov chains and polynomial time algorithms. In: 35th Annual Symposium on Foundations of Computer Science. Santa Fe, New Mexico: IEEE, 20–22 November 1994; 656–671. 24, 62, 68, 69

[74] Kannan R, Lovász L, Simonovits M. Isoperimetric inequalities for convex bodies and a localization lemma. Discrete and Comput Geom 1995;13(3–4):541–549. 24, 62, 64

[75] Kannan R, Lovász L, Simonovits M. Random walks and an $O^*(n^5)$ volume algorithm for convex bodies. Random Structures Algorithms August 1997; 11(1):1–50. 24, 62, 63, 79, 81

[76] Kannan R, Tetali P, Vempala S. Simple Markov-chain algorithms for generating bipartite graphs and tournaments (extended abstract). In: Proceedings of the Eighth Annual ACM–SIAM Symposium on Discrete Algorithms. New Orleans, Louisiana, 5–7 January 1997; 193–200. 23

[77] Karger DR. A randomized fully polynomial time approximation scheme for the all terminal network reliability problem. In: Proceedings of the Twenty-Seventh Annual ACM Symposium on the Theory of Computing. Las Vegas, Nevada, 29 May–1 June 1995; 11–17. 32

[78] Karp RM, Luby M. Monte-Carlo algorithms for enumeration and reliability problems. In: 24th Annual Symposium on Foundations of Computer Science. Tucson, Arizona: IEEE, 7–9 November 1983; 56–64. 30, 32

[79] Karp RM, Luby M, Madras N. Monte-Carlo approximation algorithms for enumeration problems. J Algorithm September 1989;10(3):429–448. 32

[80] Karzanov A, Khachiyan L. On the conductance of order Markov chains. Order 1991;8(1):7–15. 24, 105

[81] Knuth DE. Big omicron and big omega and big theta. SIGACT News 1976; 8(2):18–24. 1

[82] Kozen DC. The Design and Analysis of Algorithms. Springer-Verlag. ISBN 3-540-97687-6 or 3-877-97687-6, December 1991; lecture 26, 138–143. 30

[83] Lanford III OE. Entropy and equilibrium states in classical statistical mechanics. In: Lenard A, editor, Statistical Mechanics and Mathematical Problems, Springer-Verlag, Number 20 in Lecture Notes in Physics. 1973; 107–113. 18

[84] Levin L. Universal search problems. Problemy Peredachi Informatsii 1973; 9(3):265–266. In Russian; partial English translation in Trakhtenbrot [121]. 4

[85] Lindvall T. Lectures on the Coupling Method. Wiley Series in Probability and Mathematical Statistics. Wiley-Interscience, 1992. ISBN 0-471-54025-0. 74

[86] Linial N. Hard enumeration problems in geometry and combinatorics. SIAM J Algebraic Discrete Methods April 1986;7:331–335. 49

[87] Lovász L, Simonovits M. The mixing rate of Markov chains, an isoperimetric inequality, and computing the volume. In: 31st Annual Symposium on Foundations of Computer Science. St. Louis, Missouri: IEEE, volume I, 22–24 October 1990; 346–354. 62

[88] Lovász L, Simonovits M. Random walks in a convex body and an improved volume algorithm. Random Structures Algorithms 1993;4(4):359–412. 24, 62, 63, 64, 66, 68, 79

[89] Luby M, Randall D, Sinclair A. Markov chain algorithms for planar lattice structures (extended abstract). In: 36th Annual Symposium on Foundations of Computer Science. Milwaukee, Wisconsin: IEEE, 23–25 October 1995; 150–159. 17

[90] Luby M, Vigoda E. On the convergence rate of the Glauber dynamics for sampling weighted independent sets. (Preprint). 25, 83, 103

[91] Luby M, Vigoda E. Approximately counting up to four. In: Proceedings of the Twenty-Ninth Annual ACM Symposium on Theory of Computing. El Paso, Texas, 4–6 May 1997; 682–687. 17, 103, 104, 105, 130

[92] Matthews P. Generating a random linear extension of a partial order. Ann Probab 1991;19(3):1367–1392. 17, 105

[93] McDiarmid C. On the method of bounded differences. In: Siemons J, editor, Surveys in Combinatorics. Cambridge University Press, LMS Lecture Note Series 141, 1989; 148–188. 73

[94] Mihail M. On coupling and the approximation of the permanent. Inform Process Lett 1989;30:91–95. 17

[95] Mihail M, Winkler P. On the number of Eulerian orientations of a graph. In: Proceedings of the Third Annual ACM–SIAM Symposium on Discrete Algorithms. Orlando, Florida, 27–29 January 1992; 138–145. 49

[96] Motwani R, Raghavan P. Randomized Algorithms. Cambridge University Press, 1995. ISBN 0-521-47465-5. 1

[97] Papadimitriou CH. Computational complexity. Reading, Massachusetts: Addison-Wesley, 1994. 2, 49

[98] Peinado M, Lengauer T. Random generation of embedded graphs and an extension to Dobrushin uniqueness. In: Proceedings of the Thirtieth Annual ACM Symposium on Theory of Computing. Dallas, Texas, 24–26 May 1998; 176–185. 18

[99] Pilarski S, Kameda T. Simple bounds on the convergence rate of an ergodic Markov chain. Inform Process Lett 1993;45:81–87. 16

[100] Propp JG, Wilson DB. Exact sampling with coupled Markov chains and applications to statistical mechanics. Random Structures Algorithms 1996;9(1–2):223–252. 27, 28

[101] Propp JG, Wilson DB. How to get a perfectly random sample from a generic Markov chain and generate a random spanning tree of a directed graph. J Algorithm May 1998;27(2):170–217. 27, 28

[102] Pruesse G, Ruskey F. Generating linear extensions fast. SIAM J Comput April 1994;23(2):373–386. 105

[103] Randall D, Tetali P. Analyzing Glauber dynamics by comparison of Markov chains. In: Lucchesi CL, Moura AV, editors, Third Latin American Symposium on Theoretical Informatics, Campinas, Brazil: Springer-Verlag, volume 1380 of Lecture Notes in Computer Science. ISBN 3-540-64275-7, 1998; 292–304. 24, 103

[104] Robbins HE. A theorem on graphs, with an application to a problem of traffic control. Amer Math Monthly 1939;46:281–283. 100

[105] Roth D. On the hardness of approximate reasoning. Artificial Intelligence 1996;82:273–302. 128, 129, 130

[106] Salas J, Sokal AD. Absence of phase transition for antiferromagnetic Potts models via the Dobrushin uniqueness theorem. J Statist Phys February 1997; 86(3–4):551–579. 19, 111, 114

[107] Schaefer TJ. The complexity of satisfiability problems. In: Conference Record of the Tenth Annual ACM Symposium on Theory of Computing. San Diego, California, 1–3 May 1978; 216–226. 48

[108] Schnorr CP. Optimal algorithms for self-reducible problems. In: 3rd International Colloquium on Automata, Languages and Programming. Edinburgh. ISBN 0-85224-308-1, 20–23 July 1976; 322–337. 33, 52, 94

[109] Simon B. The Statistical Mechanics of Lattice Gases. Princeton, New Jersey: Princeton University Press, volume 1. ISBN 0-691-08779-2, 1993; chapter V. 18, 122

[110] Sinclair A. Improved bounds for mixing rates of Markov chains and multicommodity flow. Combin Probab Comput 1992;1:351–370. 9, 22, 23, 24

[111] Sinclair A. Algorithms for Random Generation and Counting: A Markov Chain Approach. Progress in Theoretical Computer Science. Cambridge, Massachusetts: Birkhäuser Boston, 1993. ISBN 0-8176-3658-7 or 3-7643-3658-7. 1, 6, 14, 49, 104, 127, 128, 130

[112] Sinclair A, Jerrum M. Approximate counting, uniform generation and rapidly mixing Markov chains. Inform and Comp 1989;82:93–133. 22, 23, 33, 56, 120

[113] Sokal A. Private Communication, January 1995. 18, 115

[114] Spearman C. 'Footrule' for measuring correlations. British J Psych July 1906; 2:89–108. 106, 137

[115] Stanley RP. Acyclic orientations of graphs. Discrete Math 1973;5:171–178. 49

[116] Stanley RP. Enumerative Combinatorics. volume 1. Wadsworth & Brooks/Cole, 1986. (2nd printing, Cambridge University Press, July 1997). 119

[117] Stockmeyer L. The complexity of approximate counting (preliminary version). In: Proceedings of the Fifteenth Annual ACM Symposium on Theory of Computing. Boston, Massachusetts, 25–27 April 1983; 118–126. 30

[118] Thorisson H. Coupling methods in probability theory. Scand J Statist 1995; 22:159–182. 18, 70

[119] Toda S. PP is as hard as the polynomial-time hierarchy. SIAM J Comput October 1991;20(5):865–877. 4

[120] Tomescu I. Sur le problème du coloriage des graphes généralisés. Compt Rend Heb Acad Sci Sér A 5 August 1968;267(6):250–252. 11, 39, 40

[121] Trakhtenbrot BA. A survey of Russian approaches to perebor (brute-force search) algorithms. Ann Hist Comput October 1984;6(4):384–400. 145

[122] Vadhan S. The Complexity of Counting. Bachelor's thesis, Mathematics and Computer Science, Harvard College, Cambridge, Massachusetts, April 1995. 104, 128, 129, 130

[123] Vadhan SP. The complexity of counting in sparse, regular, and planar graphs, May 1997. `http://theory.lcs.mit.edu/~salil/papers/sparse-regular.ps.gz`. 55, 104, 130

[124] Valiant LG. The complexity of computing the permanent. Theoret Comput Sci 1979;8:189–201. 5

[125] Valiant LG. The complexity of enumeration and reliability problems. SIAM J Comput August 1979;8(3):410–421. 5, 48, 53, 94, 102, 130

[126] Welsh DJA. Complexity: knots, colouring and counting. volume 186 of London Mathematical Society Lecture Notes. Cambridge: Cambridge University Press, 1993. 1, 2, 10

[127] Wilson DB. Personal Communication, August 1997. 108, 109

[128] Wilson DB. Mixing times of lozenge tiling and card shuffling Markov chains, August 1997. Manuscript. 106, 108, 109, 110

[129] Wilson DB. Annotated bibliography of perfectly random sampling with Markov chains. In: Aldous D, Propp J, editors, Microsurveys in Discrete Probability, American Mathematical Society, volume 41 of DIMACS Series in Discrete Mathematics and Theoretical Computer Science. 1998; 209–220. 28

[130] Wilson RJ. Introduction to Graph Theory. Longman, 4th edition, 1996. ISBN 0-5822-4993-7. 9

Index